Table of Contents

About the National Science and Technology Council

The National Science and Technology Council (NSTC) is the principal means by which the Executive Branch coordinates science and technology policy across the diverse entities that make up the Federal research and development enterprise. A primary objective of the NSTC is establishing clear national goals for Federal science and technology investments. The NSTC prepares research and development strategies that are coordinated across Federal agencies to form investment packages aimed at accomplishing multiple national goals. The work of the NSTC is organized under five committees: Environment, Natural Resources and Sustainability; Homeland and National Security; Science, Technology, Engineering, and Math (STEM) Education; Science; and Technology. Each of these committees oversees subcommittees and working groups focused on different aspects of science and technology. More information is avail-able at http://www.whitehouse.gov/ostp/nstc.

About the Office of Science and Technology Policy

The Office of Science and Technology Policy (OSTP) was established by the National Science and Technology Policy, Organization, and Priorities Act of 1976. OSTP's responsibilities include advising the President in policy formulation and budget development on questions in which science and technology are important elements; articulating the President's science and technology policy and programs; and fostering strong partnerships among Federal, state, and local governments, and the scientific communities in industry and academia. The Director of OSTP also serves as Assistant to the President for Science and Technology and manages the NSTC. More information is available at http://www.whitehouse.gov/ostp.

About the Interagency Arctic Research Policy Committee

The Arctic Research and Policy Act of 1984 (ARPA), Public Law 98-373, July 31, 1984, as amended by Public Law 101-609, November 16, 1990, provides for a comprehensive national policy dealing with national research needs and objectives in the Arctic. The ARPA establishes an Arctic Research Commission (ARC) and an Interagency Arctic Research Policy Committee (IARPC) to help implement the Act. IARPC was formally created by Executive Order 12501. Its activities have been coordinated by the National Science Foundation (NSF), with the Director of the NSF as chair. On July 22, 2010, President Obama issued a Memorandum for the Director of the OSTP making the NSTC responsible for the IARPC with the Director of the NSF remaining as chair of the committee.

About this Document

This report was developed by the IARPC which reports to the NSTC Committee on Environment, Natural Resources, and Sustainability (CENRS). This report is published by OSTP.

February 2013

MEMBERS OF CONGRESS:

I am pleased to forward the five-year Arctic research plan produced by the Interagency Arctic Research Policy Committee (IARPC). Creation of this plan is one of IARPC's responsibilities described in the Arctic Research Policy Act of 1984 (15 U.S.C. § 4108). Fourteen Federal agencies, departments, and offices collaborated to develop this plan, which calls for strong interagency communication, coordination, and collaboration within the framework of the National Science and Technology Council. The IARPC staff also consulted with collaborators in the State of Alaska, local communities, indigenous organizations, non-governmental organizations, and the academic community to ensure that the interests and needs of all stakeholders are addressed appropriately in this research plan.

The Arctic environment is undergoing rapid transition as sea and land ice diminish, with tremendous implications for natural environments, human well-being, national security, transportation, and economic development. The United States and the other Arctic nations require strong, coordinated research efforts to understand and forecast changes in the Arctic.

Toward that end, and in furtherance of goals developed by the Arctic Research Commission, this plan focuses on those research activities that would be substantially enhanced by multi-agency collaboration. Many important investigations outside the scope of this plan will continue to be conducted within individual agencies or through other interagency collaborations.

I appreciate your support as this Administration works to ensure that the Nation's research efforts in the Arctic are broadly coordinated across the full spectrum of Federal agencies and interests.

Sincerely,

John P. Holdren
Assistant to the President for Science and Technology
Director, Office of Science and Technology Policy

Chapter 1: Executive Summary

There is broad scientific consensus that rapid changes in global climate are altering ice and snow cover and affecting Arctic ecosystems, indigenous societies, and natural resources. Research is needed to increase fundamental understanding of these challenges and inform development of sound, science-based solutions.

The Interagency Arctic Research Policy Committee (IARPC) is charged with developing five-year plans for Federally sponsored research in the Arctic region. For 2013 to 2017, the IARPC, which consists of representatives from 14 Federal agencies, departments, and offices, has identified seven research areas that will inform national policy and benefit significantly from close interagency coordination. They are:

1. Sea ice and marine ecosystems;
2. Terrestrial ice and ecosystems;
3. Atmospheric studies of surface heat, energy, and mass balances;
4. Observing systems;
5. Regional climate models;
6. Adaptation tools for sustaining communities; and
7. Human health.

These research areas do not encompass all Federal Arctic research activities that will occur over the next five years. Many important investigations outside the scope of this plan will continue to be conducted within individual agencies or through other interagency collaborations.

Sea ice and marine ecosystems

Arctic marine ecosystems—driven largely by large-scale changes in sea ice—are moving to new states with the potential for short-term surprises. Over the next five years, the IARPC agencies—including the Department of Defense (DOD), Department of Energy (DOE), Department of the Interior (DOI), National Aeronautics and Space Administration (NASA), National Oceanic and Atmospheric Administration (NOAA), and National Science Foundation (NSF)—will increase capacity to study such rapid changes through the following activities:

1. Develop a framework of observations and modeling to support forecasting of sea-ice extent on seasonal to annual scales for operational and research needs;
2. Identify and study sites in the Beaufort and Chukchi Seas and the contiguous Arctic Ocean where climate feedbacks are active;
3. Complete deployment of a Distributed Biological Observatory (DBO) in the Arctic Ocean to create long-term data sets on biological, physical, and chemical variability and ecosystem response; and
4. Develop integrated ecosystem-processes research in the Beaufort and Chukchi Seas region.

Terrestrial ice and ecosystems

Ongoing changes in the terrestrial Arctic environment that result from climate change are expected to lead to further changes in global climate, or climate "feedbacks," and affect the ability of local communities to adapt. The IARPC has identified the following five priority activities to understand such climate feedbacks and terrestrial ecosystem processeses to be coordinated collaboratively by the DOE, DOI, NASA, NSF, and Smithsonian Institution (SI):

1. Perform glacial-process studies targeting specific dynamic ranges;

2. Coordinate and integrate efforts, including information delivery, that contribute to terrestrial ecosystem research;

3. Identify and study key sites where climate feedbacks are active, including permafrost, snow, hydrates, glaciers, and ice;

4. Investigate the frequency and severity of wildland fires in the Arctic and understand their impacts on vegetation and wildlife; and

5. Conduct socio-economic research to understand ecosystem services as the Arctic tundra changes with increased warming to inform plans for protecting, managing, and adapting to a fragile and changing Arctic environment.

Atmospheric studies of surface heat, energy, and mass balances

Variability in surface-air temperatures—from year-to-year or longer—tends to be larger in the Arctic than in other parts of the globe. Compared with those at low latitudes, atmospheric processes in the Arctic are influenced by unique features—such as polar night, high reflectivity of snow and ice cover, and atmospheric stability—that influence the degree to which aerosols and clouds warm or cool the region. Scientific uncertainties about these unique features must be clarified in order to more fully understand the Arctic atmosphere and its processes.

Coordinated remote-sensing and *in-situ* observations, improved representation of atmospheric processes in models, quantification of uncertainty in model outputs, and long-term observational data sets will be critical to addressing scientific uncertainties. DOE, NASA, NOAA, and NSF will collaborate on three activities to support this research area:

1. Improve understanding of short-lived climate forcers (SLCFs) and their role in Arctic amplification through satellite observations, long-term *in-situ* observations, and improved modeling;

2. Improve understanding of processes controlling formation, longevity, and physical properties of Arctic clouds, including the effects of—and sensitivities to—aerosols; and

3. Develop an integrated understanding of Arctic atmospheric processes, their impact on the surface-energy budget, and their linkages with oceanic, terrestrial, and cryospheric systems through improved satellite capabilities, ground-based observations, and representations of Arctic systems in climate and weather-prediction models.

Observing systems

Arctic change is occurring on multiple spatial and temporal scales. Over the next five years, the DOE, DOI, Environmental Protection Agency (EPA), NASA, NOAA, NSF, Office of Naval Research (DOD-ONR), and United States Coast Guard (DHS-USGC) will focus on nine activities to maintain and strengthen an integrated national and international Arctic observing system to obtain data and information from multiple scales:

1. Facilitate observing-system design for the Arctic;

2. Assess local-resident priorities with respect to climate;

3. Combine *in-situ* and remotely sensed observation of sea ice with local community and traditional knowledge;

4. Conduct long-term monitoring of key outlet glaciers and tidewater glaciers;

5. Monitor the biological and physical state of the Arctic marine environment;

6. Assess the effects of clouds and atmospheric constituents on surface radiation balance;

7. Assess the impact of terrestrial warming and permafrost thawing on the carbon cycle;

8. Improve data access; and

9. Engage indigenous observers and communities in monitoring environmental parameters.

Regional climate models

Models of Earth's climate are mathematical tools for understanding current climate processes and projecting future climate variability and change. Improved models of the Arctic region will enhance scientific understanding of processes occurring today, improve the accuracy of projections of future change, guide the design of more effective Arctic field-research campaigns, and support better-informed decision making.

The DOE, NOAA, DOI, and NSF will integrate modeling and process-science research through seven activities to improve modeling of Arctic systems and processes:

1. Inventory existing Federal Arctic modeling activities;

2. Encourage coordination to better represent Arctic processes in Earth-system models;

3. Build Arctic-region models to couple with regional and global approaches;

4. Develop models of Arctic land ice mass loss, connections to ocean and atmospheric variability, and implications for sea level;

5. Increase Arctic-model resolution to improve prediction and inform future research and observations;

6. Use model-derived insights to inform process research and vice versa; and

7. Improve understanding of the principle drivers and uncertainties of Arctic climate changes through model validation and verification.

Adaptation tools for sustaining communities

Arctic residents are adapting to new conditions created by rapid environmental change and diverse socio-economic stressors. Over the next five years, the Department of Agriculture (USDA), DOI, Department of State (DOS), EPA, NOAA, NSF, and the SI will assess the resilience and vulnerabilities of Arctic communities to the impacts of climate change. That assessment will help provide residents, community leaders, and policy makers with the knowledge needed to develop sound strategies for successful adaptation. The effort will focus on four activities:

1. In collaboration with local communities, develop methods for assessing community sustainability and resilience and determine the efficacy of current adaptation strategies and means for identifying unintended positive and negative outcomes;

2. Identify current vulnerabilities of Arctic communities and ecosystems to climate change and explore the interaction of climate vulnerabilities with socio-economic and other stressors;

3. Develop projections of future climate scenarios and demographic conditions to forecast potential strengths and weaknesses of Arctic human and ecological systems; and

4. Design new research, education, and outreach tools and processes to assist Arctic communities in language and heritage preservation and in cultural revitalization efforts.

Human health

Arctic indigenous peoples have shorter life expectancies and greater infant mortality rates than their respective national populations. In addition to higher death rates for unintentional injury and suicide, native peoples experience a high prevalence of both infectious diseases and health impacts associated with exposures to environmental pollutants, rapid economic change, and climate change.

IARPC human-health research activities planned for the next five years reflect the priorities of both the Arctic Research Commission *Goals and Objectives for Arctic Research 2009-2010* and the 2011 Arctic Health Ministers meeting held in Nuuk, Greenland. The Department of Health and Human Services, Centers for Disease Control and Prevention (HHS-CDC), EPA, Indian Health Services (HHS-IHS), National Institutes of Health (HHS-NIH), and the U.S. Arctic Research Commission will focus on four activity areas:

1. Continue to expand circumpolar surveillance and research for infectious, non-communicable diseases, trauma, injury, sanitation services, and indoor air quality to help prevent morbidity and mortality;

2. Continue interagency collaboration to monitor the impacts of climate change and environmental contaminants on human health and wildlife;

3. Continue to support investigator-initiated research in major health priority areas such as mental health including substance abuse and suicide, obesity, diabetes, and cancer; and

4. Continue to engage indigenous communities and tribal groups in research activities and projects in the Arctic.

Chapter 2: Introduction and Background

Lead authors:
Brendan P. Kelly, Office of Science and Technology Policy
C. Nikoosh Carlo, National Science Foundation

Introduction

Meeting the Nation's economic, scientific, and environmental needs in the Arctic requires research across diverse disciplines and the involvement of multiple Federal agencies. The Interagency Arctic Research Policy Committee (IARPC), which consists of principals from 14 agencies, departments, and offices across the Federal government, is charged with developing five-year plans for federally sponsored research in (and about) the Arctic region (Arctic Research Policy Act of 1984; Title I of P.L. 98-373 of July 31, 1984). Federal agencies that participate in the IARPC have diverse roles in carrying out the National Arctic Research Policy, which is articulated in National Security Presidential Directive 66/Homeland Security Presidential Directive 25. It mandates that—with respect to research—the IARPC will:

- Meet national security and homeland security needs relevant to the Arctic region;

- Protect the Arctic environment and conserve its biological resources;

- Ensure environmentally sustainable natural resource management and economic development in the region;

- Strengthen institutional cooperation among the eight Arctic nations (the United States, Canada, Denmark, Finland, Iceland, Norway, the Russian Federation, and Sweden);

- Involve the Arctic's indigenous communities in decisions that affect them; and

- Enhance scientific monitoring of and research on local, regional, and global environmental issues.

To ensure that these policy objectives are informed by the best possible science, this plan sets seven priority research areas for the next five years that stand to benefit significantly from close interagency collaboration. This plan does not include all Arctic research activities occurring across the Federal government—many important investigations will continue to be conducted within individual agencies or through other interagency collaborations. Individual agencies have described their own Arctic research priorities in a series of recent reports including:

The U.S. Arctic Research Commission's *Report on Goals & Objectives for Arctic Research 2009 – 2010* (http://www.arctic.gov/publications/2009-10_usarc_goals.html);

The U. S. Navy's *Arctic Road Map* (http://www.dtic.mil/cgi-bin/GetTRDoc?Location=U2&doc=GetTRDoc. pdf&AD=ADA516591);

The National Oceanic and Atmospheric Administration's *Arctic Vision and Strategy* (http://www.arctic. noaa.gov/docs/NOAAArctic_V_S_2011.pdf);

The U.S. Geological Survey's (Department of the Interior) *An Evaluation of the Science Needs to Inform Decisions on Outer Continental Shelf Energy Development in the Chukchi and Beaufort Seas, Alaska* (http://pubs.usgs.gov/circ/1370/pdf/circ1370.pdf); and

The *Changing Conditions in the Arctic* section of the National Ocean Policy Implementation Plan (http://www.whitehouse.gov/administration/eop/oceans/objectives).

This five-year plan also does not describe the considerable important and complementary Arctic research being conducted outside of the Federal government—by academic, state, tribal, and non-governmental researchers.

Under the auspices of the Arctic Council[1], Federal agencies also pursue Arctic research through work-ing groups and *ad hoc* task forces. This collaborative work focuses on a wide range of issues, including climate change, indigenous land use, and communications technology.

Successful implementation of this five-year research plan will require close coordination with all of the above listed efforts, the State of Alaska, indigenous organizations, academic institutions, non-governmental organizations, the Arctic Council, and international partners.

Scientists widely agree that the societal and environmental consequences of rapid environmental change are the most pressing scientific concerns in the Arctic region today. Diminishing sea-ice cover is expected to impact the global climate; diminishing ice sheets and glaciers are resulting in sea-level rise; and thawing permafrost is having impacts on both local infrastructure and the concentration of greenhouse gases in the atmosphere. Addressing these challenges will require a sharp research focus on the changing cryosphere—the world's solid-state water, including sea ice, glaciers, snow cover, and permafrost—and its effects on the physical environment, ecosystems, and communities in the Arctic and elsewhere. Key research questions include:

Sea ice and Arctic Ocean ecosystems

- At what rates will Arctic sea ice diminish over the next 100 years?
- What will be the consequences of diminishing sea ice for Arctic ecosystems and their inhabitants?
- What will be the consequences of diminishing sea ice for global climate and environments?
- How will Arctic Ocean acidity change in coming decades?
- What will be the consequences of acidification for Arctic ecosystems and their inhabitants?

Ice sheets and glaciers

- At what rates will Arctic glaciers and ice sheets diminish over the next 100 years and what processes and forcings are driving the loss?
- What will be the consequences of diminishing glaciers and ice sheets for Arctic ecosystems and their inhabitants?
- How will diminishing glaciers and ice sheets impact global climate and sea level?

---

1. The Arctic Council consists of the eight Arctic States: Canada, Denmark (including Greenland and the Faroe Islands), Finland, Iceland, Norway, Russia, Sweden, and the United States. Six international organizations representing Arctic Indigenous Peoples have permanent participant status.

Permafrost

- At what rates will Arctic permafrost diminish over the next 100 years?

- What will be the consequences of diminishing permafrost for Arctic ecosystems and their inhabitants?

- How will changes in permafrost impact the global climate system?

These key questions are being addressed by several Federal and non-Federal research efforts including:

The U.S. Global Change Research Program (http://www.globalchange.gov);

The National Ocean Policy Implementation Plan, *Changing Conditions in the Arctic* (http://www.white-house.gov/administration/eop/oceans/objectives); and

The interagency Study of Environmental Arctic Change (SEARCH) (http://www.arcus.org/search/searchscience).

This five-year plan identifies seven priority research areas where interagency cooperation will strengthen and enhance this ongoing work. They are:

- Sea ice and marine ecosystems;

- Terrestrial ice and ecosystems;

- Atmospheric studies of surface heat, energy, and mass balances;

- Observing systems;

- Regional climate models;

- Adaptation tools for sustaining communities; and

- Human health.

Background

The Arctic Research and Policy Act of 1984 (ARPA) established the United States Arctic Research Commission (USARC) to promote Arctic research and recommend a research policy for the region. It also established the IARPC to develop national Arctic research policy and a five-year implementation plan. The IARPC is chaired by the director of the (NSF) and consists of principal representatives from 14 Federal agencies, departments, and offices. The IARPC staff meet monthly and the Principals meet twice per year.

In May 2010, recognizing the increasing participation of multiple agencies in Arctic research, President Obama directed the IARPC to be chartered as a subcommittee of the National Science and Technology Committee. This report constitutes the committee's first deliverable since that change and subsequent revitalization under the leadership of both the Office of Science and Technology Policy and NSF.

The IARPC planning efforts adhere to Section 112 of the ARPA, which defines the Arctic as "all United States and foreign territory north of the Arctic Circle and all United States territory north and west of the boundary formed by the Porcupine, Yukon, and Kuskokwim Rivers (in Alaska); all contiguous seas, including the Arctic Ocean and the Beaufort, Bering, and Chukchi Seas; and the Aleutian chain." While fully accepting this definition, the Committee also emphasizes that the Arctic is part of a larger, changing global system—the boundaries of which must be considered flexible in order to properly study the Arctic's role in important global processes.

Previous work of the IARPC

Between 1987 and 2007, the IARPC produced 21 volumes of the biannual journal, *Arctic Research of the United States*, which reported Arctic research and results emerging from agencies and partners. The journal's broad vision was to support scientific and engineering research that implements national policy objectives, including:

- Protecting the Arctic environment and conserving its living resources;
- Promoting environmentally sustainable natural resource management and economic development in the region;
- Strengthening institutions for cooperation among the eight Arctic nations;
- Involving indigenous Arctic peoples in decisions that affect them;
- Enhancing scientific monitoring and research on local, regional, and environmental issues (including their assessment); and
- Meeting post-Cold-War national security and defense needs.

Development of the five-year Arctic Research Plan

This plan was developed by the IARPC staff: Jonathan Berkson, USGC; Shella Biallas, DOI; John Calder, NOAA; C. Nikoosh Carlo, NSF; Ashley Chappell, NOAA; Kathy Crane, NOAA; Richard Eckman, NASA; Wanda Ferrell, DOE; William Fitzhugh, SI; Martin O. Jeffries, DOD; Brendan P. Kelly, OSTP; Igor Krupnik, SI; Michael Kuperberg, DOE; Marya Levintova, NIH; Kim Mcgraw, DOI; Adrianna Muir, DOS; Alan Parkinson, CDC; James Partain, NOAA; Robert Sanford, NSF; Sandy Starkweather, NOAA; Simon Stephenson, NSF; Louis Tupas, DOA; Taneil Uttal, NOAA; Thomas Wagner, NASA. The Arctic Research Commission also provided valuable input to this plan.

Evolution of scientific studies in the Arctic

Early scientific information about the Arctic was primarily geographic and collected by 18th and 19th century European explorers (Beechey 1831; Hall 1866). For most of the following 100 years, scientific information came mainly from scattered efforts in ethnography and natural history, mostly associated with expeditions by the U.S. Signal Corps and others (Dall 1870; Ray 1885) and with management of fishing and hunting (Allen 1880; Elliot 1898).

The first intensive investigation along the Arctic coast of Alaska took place with the International Polar Year (1881–1884) when the Signal Corps occupied a research station at Barrow, Alaska (Baker 1982). In the 1940s, ONR established the Naval Arctic Research Laboratory at Barrow, and studies of Arctic environments have been carried out there almost continuously under the administration of the Navy, the University of Alaska, and the North Slope Borough.

In the late 1950s, the Federal government considered detonating nuclear devices to create a port along the Chukchi coast of Alaska (AEC 1959). The Atomic Energy Commission contracted an investigation of potential environmental impacts and, thereby, provided baseline information about the Chukchi coast of Alaska and its near-shore waters (Willimousky and Wolfe 1966). In the 1970s and 1980s, an interagency agreement between the DOI's Bureau of Land Management and the Department of Commerce (DOC) NOAA created the Outer Continental Shelf Environmental Assessment Program to study the potential impacts of offshore oil development in sub-Arctic and Arctic Alaskan waters. For a decade or more, hundreds of studies looked at ice movements and deformation, mammals, birds, fish, benthos, plankton, microbiology, chemistry, oceanography, meteorology, and geology. Some of the work was eventually published in peer-reviewed literature, and all of it has been assembled by the Alaska Resources Library and Information Services housed on University of Alaska Anchorage campus (http://www.arlis.org/resources/ocseap-reports).

Through the 1980s, most Arctic research was conducted in isolated disciplines such as biology, geology, and anthropology. Toward the end of the decade, systems science matured (Ashby 1956; von Bertalanffy 1972; Lawton 2001), and a realization that human activity was driving rapid change in the Arctic began to prompt interdisciplinary work. The importance of interactions among systems—atmosphere, geosphere, hydrosphere, and biosphere—was underscored by the International Geosphere-Biosphere Programme and the World Climate Research Programme. The notion also stimulated the formation of NSF's Arctic System Science Program, with a research focus on paleoenvironments and contemporary studies of interactions among ocean, land, and atmosphere. Eventually, human dimensions were included in studies of contemporary and paleoenvironments.

While the IARPC's mission is focused on the Arctic region, many new systems science approaches have been developed to consider high latitude phenomena (especially those associated with ice sheets, sea ice, and atmospheric coupling) in a broader perspective. As a result, there have been many commonalities in the evolution of Arctic and Antarctic science.

In 1997, scientists from 25 institutions called for a coordinated effort to understand rapid environmental change in the Arctic. Their efforts led to the formation of the SEARCH (http://www.arcus.org/search/index.php). They produced science and research plans based on three main components: observing, understanding, and responding to Arctic change. The Arctic Observing Network (AON; http://www.arcus.org/search/aon)—a component of SEARCH—aims to track and foster understanding of the complex, rapid environmental changes taking place in the Arctic through modeling, reconstructions of paleoen-vironments, and process studies of the environment, socio-economics, cultures, and human health. Responding to change requires consideration of possible adaptive responses of Arctic communities and possible effects on people living outside the Arctic region. Rapid warming of the Arctic has led to dramatic declines in sea-ice extent and thickness with local and global impacts. Arctic sea ice influences atmospheric circulation patterns and precipitation as far south as the tropics (Budikova 2009). Similarly, warming has led to substantial losses of land ice, primarily from Greenland and Alaska, which are now responsible for a substantial fraction of observed sea level rise.

Summary

Interagency cooperation on Arctic research is more important than ever. Rapid changes are affecting the region's biota and people in many ways, including by increasing access to the region for energy and mineral development, shipping, tourism, and military operations—human activities that may carry both risks and opportunities for the Arctic region. Federal agencies are conducting scientific research to fundamentally understand those changes, risks, and opportunities. Policymakers are increasingly relying on that science to make decisions and form practical responses. This IARPC research plan aims to support those decisions with enhanced interagency cooperation on Arctic research to address the most pressing science needs.

References

Allen, J. S. (1880). *History of North American pinnipeds; a monograph of the walruses, sea-lions, sea-bears and seals of North America.* Washington: Government Printing Press.

Ashby, W. R. (1956). *Introduction to Cybernetics.* New York: Wiley.

Atomic Energy Commission (AEC). (1959), *Roads and Core Holes: Project Chariot Site Plan.* Holmes and Narver, Inc.

Baker, F. W. G. (1982). The First International Polar Year, 1882-1883. *Polar Record*, 21(132), 275-285.

Beechey, F. W. (1831). *Narrative of a voyage to the Pacific and Beering's Strait.* London: Henry Colburn and Richard Bently.

Budikova, D. (2009). Role of Arctic sea ice in global atmospheric circulation: A review. *Global and Planetary Change* 68, 149-163.

Dall, W. H. (1870). *Alaska and its resources.* Boston. Lee and Shepard.

Elliot, H. W. (1898). Report on the seal islands of Alaska. House Exec. Doc. 92, pt. 3-1.

Hall, C. F. (1866). *Arctic researches and life among the Esquimuax: being the narrative of and expedition in search of Sir John Franklin, in the years 1860, 1861, and 1862.* New York, NY: Harper and Bros., p. 565

Lawton, J. (2001). Earth System Science. *Science* 292:1965.

Ray, P. H. (Ed.). (1885). Report of the International Polar Expedition to Point Barrow, Alaska. 48th Congress, 2nd Session, House Exec. Doc. 44.

Steller, G. W. (1988). *Journal of a Voyage with Bering 1741-1742* O. W. Frost (Ed.), Stanford University Press.

von Bertalannfy, L. (1972). The history and status of general systems theory. *The Academy of Management Journal* 15, 407-426.

Wilimousky, N. J., & J. N. Wolfe. (1966). Environment of the Cape Thompson Region. Alaska: U.S. Atomic Energy Commission, Washington, D.C.

Chapter 3: Research Areas

3.1. Understand sea-ice processes, ecosystem processes, ecosystem services, and climate feedbacks in the Beaufort and Chukchi Seas and the contiguous Arctic Ocean

Lead Author:
John Calder, NOAA

Agency Partners:
DOD, DOI, NASA, NOAA, NSF

The annual Arctic Report Card (http://www.arctic.noaa.gov/reportcard) integrates the latest information on the state of the Arctic based on input from more than 100 authors around the world. The 2011 Report Card concludes that data collected since 2006 are sufficient to indicate a shift in the Arctic Ocean system that is characterized by the persistent decline in the thickness and summer extent of sea-ice cover and a warmer, fresher, and more acidic upper ocean. There is, moreover, growing evidence that those changes are forcing marine ecosystems in the Beaufort and Chukchi Seas and the contiguous Arctic Ocean toward new and generally unknown states, with the potential for short-term surprises (Krupnik and Bogoslovskaya 1999; Grebmeier 2006; Gradinger, *et al.* 2010).

The IARPC agencies will work together over the next five years to enhance understanding of changing sea-ice-ocean-atmosphere interactions and feedbacks; improve sea-ice forecasting at various spatial and temporal scales; and detect and understand ecosystem change. Both "understanding" and "fore-casting" involve the application of models, whose outputs will inform decision making and planning for future observations. In conjunction with work described in other sections of this plan (e.g., section 3.2), agency partners will enhance efforts to organize, disseminate, and analyze relevant data and informa-tion. Applying modern cyberinfrastructure will improve capabilities for integrating data from different sources and blending different types of data to produce new insights and information. These advances are needed to be as responsive as possible to the needs of local residents, businesses, government decision-makers, and managers of Arctic Ocean resources.

The work focuses on four activities.

3.1.1. Develop a framework of observations and modeling to support forecasting of sea-ice extent on seasonal to annual scales for operational and research needs

Why do this?

The Arctic Ocean system has shifted to a new state. Evidence cited in the Arctic Report Card shows that the minimum extent of summer Arctic sea ice from 2007 to the present has fallen below the previously established trend line for the period 1979 through 2006, and the rate of sea-ice loss exceeds that projected by coupled climate models (Stroeve *et al.* 2007)[2].

Continued loss of sea ice will have important consequences for marine and terrestrial ecosystems, coastal communities, maritime transportation, natural resource development, regional and global weather and

---

2. As this report was being finalized, a new record minimum was recorded for the extent of Arctic sea ice (http://nsidc.org/arcticseaicenews/).

climate, and national security. Understanding and predicting the consequences of continuing sea-ice loss on the marine ecosystem will require better understanding of the Arctic Ocean environment and processes to improve sea-ice forecasts and predictions at a variety of spatial and temporal scales.

Timeframe:

Mid-term (3-5 years)

Expected Outcomes:

- Improved operational and research forecasts/projections to support safe operations and eco-system stewardship on a seasonal to annual basis.

- Reduce uncertainty in predictions and projections at longer time scales and over the entire Arctic marine area, for better informed policy and decision making at local, state, national, and international levels.

- Significantly improved seasonal weather and sea-ice models and forecasts to fill a critical gap in marine weather and climate services, since at seasonal and shorter time scales, sea-ice and weather forecasts are tightly coupled and must be pursued together. Improved models and forecasts will benefit community and subsistence activities, management of protected marine resources (including ice-dependent species), marine navigation, and industry operations.

- At annual and longer time scales, reduced uncertainty and increased accuracy of sea-ice projections and enhanced understanding of how newly sea-ice-free areas influence weather and climate, not just in the Arctic but in the global system. Enhanced understanding will lead to sustainable infrastructure and community planning and aid in projection of regional and global climate impacts forced by changes in the Arctic.

Milestones:

- Convene interagency expert group on sea-ice forecasting to develop multi-year implementation plan, coordinate on-going observation and modeling, and determine needed improvements to reduce uncertainty in forecasts [DOI's Bureau of Ocean Energy Management (BOEM), NASA, NOAA, NSF, ONR; FY2013][3].

- Engage with stakeholders and users to determine needs for sea-ice forecasts and products through venues such as the Alaska Marine Science Symposium and Alaska Forum on the Environment (BOEM, NOAA, NSF, ONR; FY2013).

- Continue the Sea Ice Outlook and Sea Ice for Walrus Outlook to evaluate diverse sea-ice fore-casting approaches and fill a valuable user need (NSF, NOAA; FY2013).

- Launch Ice, Cloud, and land Elevation Satellite (ICESat) 2 satellite altimetry mission to continue the record of sea-ice thickness measurements and land ice elevation change (NASA; FY2016).

- Launch Gravity Recovery and Climate Experiment (GRACE) Follow-On satellite mission to continue the record of changes in Arctic Ocean circulation and land ice mass loss (NASA; FY2017).

---

3. Throughout this document, agencies are listed alphabetically in milestones, and that order does not imply priority.

- Develop algorithms for the Advanced Microwave Scanning Radiometer 2 (AMSR2), recently launched on Japan's GCOM-W, to continue and enhance passive microwave record of sea-ice extent (NASA, NOAA; FY2014).

- Improve knowledge of sea-ice melting through various activities such as ONR's marginal ice zone program and NASA's Operation IceBridge mission (NASA, NOAA, NSF, ONR; FY2017).

Science and Technology Gaps and Needs:

Enhanced environmental observations and improved modeling capabilities are needed to meet the requirements for improved sea-ice forecasts on a seasonal to annual basis. Many of the observations and modeling capabilities would also be valuable for sea-ice forecasts at sub-seasonal time scales. Appropriate collaborations are needed to ensure data and model outputs are available to support forecasting at both time scales.

Both *in-situ* and remotely sensed observations will be needed, taking full advantage of international remote sensing assets. Processes requiring assessment include ice growth, export, melt, and albedo change, with *in-situ* measurements coupled to remotely sensed observations to create a pan-Arctic set of sea-ice state variables for data assimilation and model initialization.

Specific sea-ice characteristics that need to be assessed are ice concentration, ice thickness, snow thickness, ice type (first-year vs. perennial), ice motion, leads/polynyas, melt pond fraction, surface albedo, temperature, and bottom and top ablation. Ocean mixed layer temperature, tidal, bathymetric, and circulation data are also needed. Most of these data also have applications to ecosystem and coastal zone studies and would lead to improved understanding of storm surges.

Planning for observations must include input from modeling centers to ensure data collection in a manner appropriate to needs for initialization, validation, and assimilation of various forecast models. Continuous or frequently repeated data collection will be needed, including broad surveys of ice conditions in at least spring and fall, to initialize forecasts of both ice loss and regrowth. To meet the observational requirements in a cost-effective way, it will be necessary to take full advantage of all available observing platforms (e.g., ships, aircraft, fixed offshore platforms, coastal locations) on an opportunistic basis. Partnerships with national and international organizations and with private industry are needed so that platforms can be equipped with instrumentation for many of the needed observations on a mutually beneficial basis. Whenever possible, data should be returned in near real-time to support forecasting at shorter time scales and to verify sensor performance.

International collaboration will be a necessary component of this project, not only because Canada and Russia share the target region with the United States, but also because international collaboration is needed for access to critical remote sensing data. The most important of these are the international satellite radar missions, which are important for constraining sea-ice age, thickness, and motion [e.g., European Space Agency (ESA) CryoSat-2 and Canadian Search And Rescue Satellite Aided Tracking (SARSAT) Missions].

3.1.2. Identify and study sites in the Beaufort and Chukchi Seas and the contiguous Arctic Ocean where climate feedbacks are active

Why do this?

There are several feedback processes active in the Arctic marine environment, including those related to albedo and radiative-balance changes and air-sea fluxes of heat, moisture, and greenhouse gases. The loss of sea ice can affect all of these processes. The extent of summer Arctic sea ice is declining, the ratio of single-year to multi-year ice continues to increase, and the rate of sea-ice loss exceeds that projected by coupled climate models (Stroeve *et al.* 2007).

The sea-ice edge during the Arctic spring through fall seasons is dynamic with rapid and large location changes in response to amount of snow cover; clouds, solar radiation, and albedo; winds and ocean waves; and air and water temperatures. Interactions and feedbacks among those variables, from local to regional scales, can amplify Arctic-wide climate change and sea-ice retreat (Perovich and Richter-Menge 2009). Better knowledge about such feedbacks is vital for understanding ice-air-ocean system processes, improving daily to seasonal weather and sea-ice forecast models, and increasing the accuracy of longer-term sea-ice and climate projections.

Changes in sea ice and other portions of the cryosphere in the Beaufort and Chukchi Seas region may affect fluxes of greenhouse gases. Recent work on the Beaufort Sea shelf found no conclusive evidence of massive methane venting from subsurface systems (Coffin *et al.* 2010), and there is no evidence in high latitude atmospheric greenhouse gas concentration data that natural emissions of methane have increased significantly in the last decade (Dlugokencky and Bruhwiler 2012). On the other hand, recent evidence (Shakova *et al.* 2010) indicates that methane is being released from or through thawing permafrost under shallow coastal seas north of Eurasia. The amount of methane available for release is potentially very large, but there are no data on the current rate of release throughout the Arctic or how the rate might change. The shallow shelves off northern Siberia and around the McKenzie Delta are important marine areas for methane research, and partnerships with Russia and Canada are essential to this work.

Carbon dioxide fluxes between the atmosphere and ocean may change with a persistent loss of sea-ice cover and warming of the Arctic Ocean. Carbon dioxide may be increasing or decreasing in the ocean at varying rates (Bates *et al.* 2011), and predicting the future influence of these fluxes on atmospheric carbon dioxide levels and on ocean acidification is not possible at present.

Timeframe:

Mid-term (3-5 years)

Expected Outcomes:

- Improved understanding of the role of feedback processes driving sea-ice variability and change; improved sea-ice forecasts to support safe operations, ecosystem stewardship, and improved predictions and projections of regional and global climate; better informed policy and decision making at local, state, national, and international levels.

- Improved daily, weekly, and seasonal weather and sea-ice models and forecasts will fill a critical gap in marine weather and climate services with benefit for community and subsistence activities, management of protected marine resources (including ice-dependent species), marine navigation, and industry operations.

- Reduced uncertainty and increased accuracy of sea-ice projections and understanding of how newly sea-ice-free areas influence weather and climate, which will lead to sustainable infrastructure and community planning and aid in projecting of regional and global climate impacts forced by changes in the Arctic.

- Improved understanding and ability to forecast (based on both *in-situ* and remotely sensed data) Arctic environmental change.

- Improved estimates of the contribution of Arctic greenhouse gas fluxes to climate warming.

Milestones:

- Report of a workshop on sea-ice forecasting (NOAA; FY2012[4]).

- Report of a workshop on the future of Arctic sea-ice research and forecasting, National Academy Polar Research Board (NASA, ONR; FY2013).

- Work through the Oil Spill Recovery Institute (OSRI) in conjunction with State of Alaska and response organizations to apply research results to oil spill response planning (DOI, NOAA, USCG[5]; FY2013).

- Investigate the marginal ice zone:

  (1) Emerging Dynamics of the Marginal Ice Zone (ONR; FY2016).

  - Science and field experiment planning; equipment development and testing (ONR; FY2013).

  - Main field experiment in the Beaufort Sea (ONR; FY2014).

  - Data analysis and synthesis (ONR; FY2016).

  (2) Marginal Ice Zone Observations and Processes Experiment (MIZOPEX): (NASA, NOAA; FY2013).

- Develop and test large-class, heavy payload unpiloted aerial systems (UAS) for sea-ice characterization (NASA, NOAA; FY2013-14).

- Investigate sea-state and boundary layer physics in the emerging Arctic Ocean (ONR; FY2017).

  - Science and field experiment planning; equipment development and testing (ONR; FY2014).

  - Main field experiment in the Beaufort and Chukchi seas (ONR; FY2015).

  - Data analysis and synthesis (ONR; FY2017).

---

4. While this plan covers the period 2013 to 2017, some activities extend work in progress. Some milestones may have been reached in 2012.
5. United States Coast Guard (Department of Homeland Security)

- Initiate interagency activity to improve application of remote sensing and buoy/mooring data to sea-ice forecasting (NOAA; FY2012).

- Investigate characterization of the circulation on the continental shelf areas of the Northeast Chukchi and Western Beaufort Seas (BOEM; FY2012).

- Continue operation IceBridge acquisition of sea-ice surface elevation and supporting data, and expand Arctic sea-ice observations to constrain melting processes (NASA; FY2017).

- Initiate interagency evaluation of trends and significance of methane flux to the atmosphere in Arctic regions (DOE, NOAA, USGS[6]; FY2015).

- Initiate a dialogue with Roshydromet and Russian Academy of Science on a potential investigation of the current rate of methane release from the shallow shelves off northern Siberia (NOAA; FY2013).

- Initiate a dialogue with Canadian agencies on collaborative methane research along the Beaufort Sea coast (DOE; FY2013).

- Identify optimal sites for short-term process studies underpinned by long-term observations (all agencies; FY2014).

Science and Technology Gaps and Needs:

New and enhanced communications technology for *in-situ* orbital and surface-based sensors is needed to provide real-time, integrated observations and products derived from atmosphere, ice, and ocean. The sensors will support investigations of the interactions and feedbacks among variables that control sea-ice concentration, thickness and motion, and the location of the ice edge. Deployment of *in-situ* sensors and support for process studies will require coordination with and access to charter and non-charter vessels capable of operating during spring, summer, and fall. Research into interactions and feedbacks among the atmosphere, ice, and ocean requires improved process models. Improving forecasts and predictions will require models that assimilate advanced observing data and derived products; provide time-varying sea-ice concentration, thickness, and ice-edge location at high temporal and spatial resolution; and fully couple ice-ocean-atmosphere processes. Improved sea-ice predictions also are critical to meeting objectives described in the following parts of this section requiring new science integration efforts on an interdisciplinary basis.

3.1.3 Complete deployment of a Distributed Biological Observatory in the Arctic Ocean to create long-term data sets on biological, physical, and chemical variability, change, and ecosystem response

Why do this?

Changes in location and timing of the seasonal ice edge can have profound effects on sea floor and water column marine ecology and human activity (Grebmeier *et al.* 2012; Huntington 2009). These changes also affect the ability of ice-dependent marine mammals to reproduce and rear young on ice (Kelly 2001;

---

6. United States Geological Survey (Department of Interior)

Kelly *et al.* 2010; Cameron *et al.* 2010; Hezel *et al.* 2012). Changes in zooplankton availability can affect distribution and abundance of baleen whales, which are important to subsistence cultures (Ashjian *et al.* 2010; Moore *et al.* 2010). Likewise, stranding of ice-dependent species on land likely reduces their survival or reproductive rates and will change their availability to subsistence hunters. Relationships among ice-edge retreat, changes in plankton dynamics, loss of summer sea ice, and foraging success of whales and ice-dependent species are poorly understood (Moore and Huntington 2008; Kovacs *et al.* 2011), as are the effects of these changes on Alaska Natives who depend upon such species (Metcalf and Robards 2008). In the initial study region (Fig. 3.1.1) changes in ecosystems will be partly driven by the varying flux of mass, heat, salt, and nutrients through the Bering Strait as well as by sea-ice loss and other changes in the physical state of the Arctic Ocean. Also, the effects on marine ecosystems of ocean acidification, happening now and projected to be greater in Arctic waters than anywhere else, are largely unknown.

The Distributed Biological Observatory (DBO) (http://www.arctic.noaa.gov/dbo) integrates biological and physical sampling from both moorings and ships using a collaborative network of logistical sup-port (Grebmeier *et al.* 2010). Remote sensing and advanced *in-situ* technologies will be critical to the observatory's success. Information from the DBO will provide a better understanding of how climate change affects Arctic biology and what steps will be necessary to improve stewardship of the Arctic marine ecosystem as human use and economic development increase. The DBO is intended to span multiple decades and likely will evolve as new approaches and technologies become available.

Figure 3.1.1. Five possible regional locations of Distributed Biological Observatory transect lines and stations for standard hydrological and biological measurements in the Pacific Arctic sector. These locations were selected because they are known biological "hot spots" where change might be easier to detect.

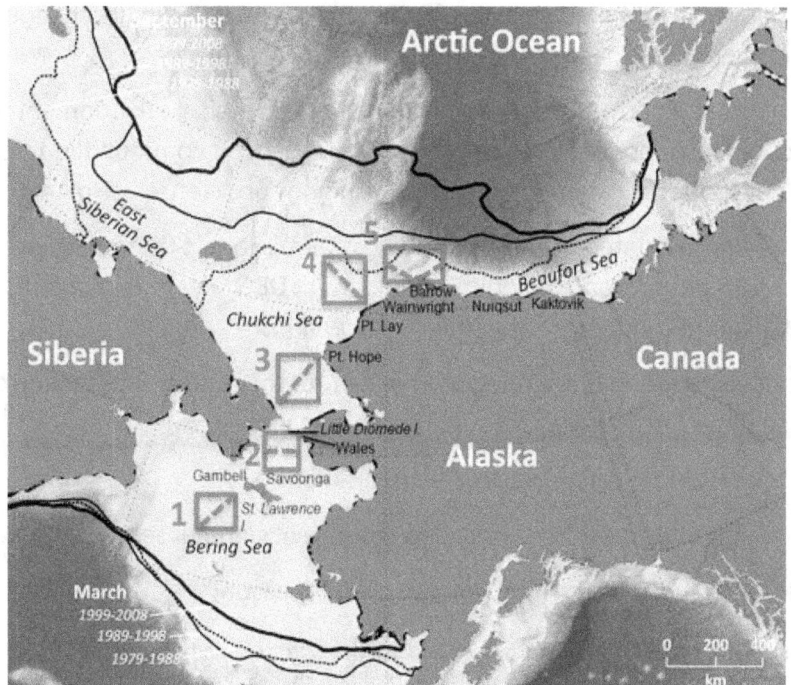

Relevant U.S. agencies (including NOAA, NSF, and NASA) coordinated through the IARPC and inter-national partners (including Canada, China, Japan, Korea, and Russia) coordinated through the Pacific Arctic Group (PAG)[7] will take advantage of current research programs to create a marine DBO in the region for consistent, long-term monitoring of biophysical responses in pivotal oceanographic areas along a north-south latitude (Figure 3.1.1). Each area exhibits high biological productivity, biodiversity, gradients in ecosystem properties, and direct linkages to subsistence-based coastal communi-ties. All areas are projected to experience increased commercial use with the loss of sea ice. As sea ice retreats, the DBO will track the rate of ecosystem change and identify impacts. The DBO will provide critical information on the biodiversity of this region and a baseline for assessing how biodiversity will respond to climate change and loss of sea ice. The DBO will also provide baseline information necessary to assess and mitigate potential impacts of offshore resource development on subsistence activities. Observations will be conducted in collaboration with international partners and coordi-nated through the Pacific Arctic Group, which is a forum for discussion of science issues relevant to the Pacific sector of the Arctic. Its participants are drawn primarily from Canada, China, Japan, Korea, Russia, and the United States.

Timeframe:

Pilot-project and planning (3-5 years); decadal-term implementation

Expected Outcomes:

The DBO will:

- Provide a knowledge-resource base to improve the ability of resource management agencies (e.g., BOEM and NOAA) to determine the effects of their actions on marine resources, resulting in improved conservation, protection, and management of Arctic coastal and ocean resources.

- Improve understanding of the effects of Arctic ecosystem and climate changes on subsistence cultures in the region.

- Provide information on variability and change in the physical environment at DBO sites and the associated response of planktonic, pelagic, and benthic communities to enable improved management and use of marine resources, including subsistence use of marine mammals.

- Develop or improve integrated physical-ecosystem models that relate past variability and change in both physical and biological conditions at DBO sites and inform resource managers of potential future conditions under different scenarios.

- Aid in developing risk-averse strategies to maximize the resilience of marine ecosystems, and develop strategies to mitigate and adapt to adverse impacts.

- Increase understanding of the ecological implications of:
  - Increasingly early ice-edge retreat and absence of summer sea ice;
  - Increased severity of storms during the ice-free season;

---

7. The PAG is an informal association of organizations and scientists from Canada, China, Japan, Korea, Russia, and the United States who coordinate ship-based observations and facilitate shared operations in the Pacific sector of the Arctic and work toward data exchange and synthesis.

- – Ice-dependent species forced to spend time on land, including impacts of human disturbance; and

- – Ocean acidification on Arctic marine ecosystems, especially plankton and calcareous sea-floor dwelling organisms important as prey to subsistence species.

- Increase ability to monitor and assess environmental conditions under changing climate scenarios through new collaborations and partnerships among participants of the Pacific Arctic Group and across U.S. agencies.

- Forge connections between a U.S. observing network and similar networks of other countries.

Milestones:

- DBO partners conduct pilot research cruises (http://www.arctic.noaa.gov/dbo/) (NOAA, NSF; FY2013).

- Creation of DBO satellite ocean color record (NASA; FY2014).

- The Arctic Observing Network (AON) subcommittee established by the IARPC organizes the DBO interagency working group to develop U.S. plans and priorities (NASA, NOAA, NSF; FY2012).

- Initiate a dialogue with *Roshydromet* and Russian Academy of Science on developing DBO stations in Russian territory as a complement to those in U.S. waters (NOAA; FY2013).

- Pacific Arctic Group meets annually to review results from 2010-2013 pilot activities (NOAA; FY2013).

- Report in 2014 on International DBO activities and results to date (NOAA; FY2014).

- Updated DBO concept and national/international plan for decadal-scale implementation release in 2014 will include identification of satellite resources that will be critical to the DBO. Ocean color, sea surface temperature (SST), sea surface height (SSH), sea surface (SS) salinity, and winds are all key measurements in the cloudy Arctic for ecosystem characterization (NASA, NOAA, NSF; FY2014).

- Starting in 2015, DBO partners execute decadal-scale plans and prepare periodic assessments on physical and ecological state of Pacific Arctic marine environment using not only DBO data, but also data from BOEM, the North Pacific Research Board, and other sources (BOEM, NOAA, NSF; FY2015).

- Report annually on recent developments at the Alaska Marine Science Symposium and seek coordination with Alaska state agencies, oil industry, and other non-Federal organizations. (Selected agencies as appropriate; FY2014).

Science and Technology Gaps and Needs:

The DBO will require agencies to pursue new technologies for continuous, year-round, real-time observations of key physical, chemical, and biological variables; improve coordination with and access to charter and non-charter vessels capable of working in Arctic areas during the spring, summer, and fall; enhance application of satellite- and aircraft-based sensors (including use of unmanned aircraft); and improve use of community based observations and instrumented animals.

3.1.4 Develop integrated ecosystem research in the Beaufort and Chukchi Seas

Why do this?

Research is needed to identify and understand processes that control ecosystem structure and function as well as their sensitivities to changes in physical and chemical environments. Results from such process studies inform models that project future ecosystem status and provide critical information to support adaptation efforts. Coordination of the Bering Sea Ecosystem Study (NSF) and the Bering Sea Integrated Ecosystem Research Program (North Pacific Research Board) stands as a model for how an interagency study might be organized (Bering Sea Interagency Working Group (2006).

There are numerous outstanding research questions that will benefit from an interagency approach, such as:

- How will ecosystems respond to expected continued warming and acidification of the waters in the Chukchi and Beaufort seas and the contiguous Arctic Ocean?

- How might fluxes of mass, heat, salt, and nutrients from the Bering Strait, from the Mackenzie River, from Siberian coastal currents, or from Atlantic water intrusion impact ecosystems in the study region?

- How will ice-edge ecosystems adapt to changes in location and timing of sea-ice retreat and re-growth?

- Will the trend toward increased areas of open water seen in recent summers lead to seasonal deepening of the mixed layer and alter ecosystem processes?

- Will southern species establish themselves in the Arctic Ocean and might the resulting stocks approach commercial size?

- Will eastern and western species and stocks intermix, and will intermixture decrease diversity and/or adaptation?

- How well do ecosystem models describe the current state of ecosystems in the study region, and what improvements are needed to enable skillful projection of future ecosystem states?

- What new observing tools or technologies are needed to improve understanding of ecosystem processes?

- How will degradation of submerged permafrost and natural gas hydrates impact ecosystems in the Chukchi and Beaufort shelf regions?

- How will increased oil and gas drilling in the Chukchi and Beaufort seas impact natural ecosystems?

- How will shipping traffic and other human activities impact subsistence species and the communities that depend upon them?

Timeframe:

Mid-term (3-5 years)

Expected Outcomes:

This research will create new knowledge about the regional ecosystem, including its key components and their linkages, and insight into how it responds to perturbation. This information will help in the development of hypotheses about responses to long-term trends such as persistent loss of summer sea ice, ocean warming, and ocean acidification. Scenarios for future subsistence and commercial use of living marine resources could be constructed from the research results and serve as guides for longer-term sustained observations. Results from process studies inform models that can be used to project future ecosystem status and allow proactive adaptation to changing conditions.

Milestones:

- Conduct interagency and international workshops and consultations during 2013 to identify high priority research themes and objectives and coordinate funding and logistic plans (Selected agencies as appropriate; FY2013).

- Perform synthesis and assessment during 2013-2014 on existing data and information to provide foundation for new research activities (Selected agencies as appropriate; FY2014).

- Initiate 3-5 year research activities starting in 2014 with interagency/international results integration mechanism (Selected agencies as appropriate; FY2014).

- Demonstrate new and updated cyberinfrastructure tools to enhance data integration and application and identify opportunities for sharing of technology and tools among interagency partners (all agencies; FY2013).

- Conduct initial science integration conference in 2016 (Selected agencies as appropriate; FY2016).

- Conduct environmental and integrated risk assessments to evaluate the potential impacts of oil/natural gas production on ecosystems in the Beaufort Sea (DOE; FY2014).

- Evaluate ecosystem impacts of oil and gas development and potential oil spills (including effects of oil dispersants and a mixture of oil and dispersants) on the early life-history stages of Arctic cod (NOAA; FY2013).

Science and Technology Gaps and Needs:

Improved partnerships between scientists and local-knowledge holders are needed as are the means for reviewing and funding such partnerships. A number of successful projects have been supported that encourage or require such partnerships, but there is recognition that more needs to be done to capture local knowledge in the Arctic. Agencies have differing mechanisms for supporting these types of partnerships but are open to good ideas generated at the working level (local residents, academic experts, and agency scientists) and proposed for consideration. The planned Arctic Observing Summit in 2013 is one opportunity for such ideas to be presented. Improved biophysical models will be needed, along with a parallel effort to test and validate the model outputs. Full-year investigations will create new requirements for observing platforms, data storage, and transmission.

References

Ashjian, C. J., Braund S. R., *et al.* (2010). Climate variability, oceanography, bowhead whale distribution and Inupiat subsistence whaling near Barrow, Alaska. *Arctic* 63(2), 179-194.

Bates, N. R., Cai W-J., & Mathis, J. T. (2011). The ocean carbon cycle in the western Arctic Ocean: Distribution and air sea fluxes of carbon dioxide. *Oceanography* 24(3), 186-201.

Bering Sea Interagency Working Group. (2006). Climate change and the Bering Sea ecosystem: An integrated, interagency, multi-institutional approach.

Cameron, M. F., Bengtson, J. L., Boveng, P. L., Jansen, J. K., Kelly, J. K., Dahle, S. P., Logerwell, E. A., Overland, J. E., Sabine, C. L., Waring, G. T., & Wilder, J. M. (2010). Status review of the bearded seal (*Erignathus barbatus*). U.S. Dep. Commer., NOAA Tech. Memo. NMFS-AFSC-211, 246 p.

Coffin, R., *et al.* (2010). First Trans-shelf-slope Climate Study in the U.S. Beaufort Sea Completed. *FITI Newsletter.*

Dlugokencky, E. J., & Bruhwiler, L. (2012). Sources of atmospheric carbon dioxide and methane in the Arctic. *Bulletin of American Meteorological Society*, 93(7), S144.

Grebmeier, J. M. (2012). Shifting patterns of life in the Pacific Arctic and Sub-Arctic Seas. *Annual Review of Marine Science.* Rev. Mar. Sci. 4(16.1-16.6).

Grebmeier, J. M., *et al.* (2010). Biological response to recent Pacific Arctic sea ice retreats. *EOS*, 91(18), 161-168.

Hezel, P.J., X. Zhang, C.M. Bitz, B. Kelly, and F. Massonnet (2012). Projected decline in snow depth on Arctic sea ice caused by progressively later autumn open ocean freeze-up this century, Geophysical Research Letters, VOL. 39, L17505, doi:10.1029/2012GL052794.

Huntington, H. P. (2009). A preliminary assessment of threats to arctic marine mammals and their conservation in the coming decades. *Mar. Policy* 33: 77-82.

Kelly, B. P. (2001). Climate change and ice breeding pinnipeds. In G. R. Walther, C. A. Burga, and P. J. Edwards (Eds.). *"Fingerprints" of Climate Change—Adapted Behavior and Shifting Species Ranges* (pp. 43-55). New York, NY, Kluwer Academic/ Plenum Publishers.

Kelly, B. P., Bengtson, J. L., Boveng, P. L., Cameron, M. F., Dahle, S. P., Jansen, J. K., Logerwell, E. A., Overland, J. E., Sabine, C. L., Waring, G. T., & Wilder, J. M. (2010). Status Review of the Ringed Seal (Phoca *hispida*). NOAA Tech. Memo. NMFS-AFSC-212.

Kovacs, K. M., Lyderson, C., *et al.* (2011). Impacts of changing sea ice conditions on Arctic marine mammals. *Marine Biodiversity.* 41, 181-194.

Metcalf, V., & Robards, M. (2008). Sustaining a healthy human-walrus relationship in a dynamic environ-ment: challenge for comanagement. *Ecol App.* 18(2), S148-S156.

Moore, S. E., & George, J. C., *et al.* (2010). Bowhead whale distribution and feeding near Barrow, Alaska in late summer 2005-06. *Arctic* 63(2), 195-205.

Moore, S. E., & Huntington, H. P. (2008). Arctic marine mammals and climate change: impacts and resilience. *Ecol App.* 18(2), S157-S165.

NSIDC. http://nsidc.org/data/seaice_index/. Last accessed, Jan. 22. 2013.

Perovich, D. K., & Richter-Menge, J. (2009). Loss of sea ice in the Arctic. *Annual Review of Marine Science*, 1, 417-441.

Shakova, N., Semiletov, I., *et al.* (2010). Extensive Methane Venting to the Atmosphere from Sediments of the East Siberian Arctic Shelf. *Science*, 327, 1246-1250.

Stroeve, J., Holland, M. M., Meier, W., Scambos, T., & Serreze, M. (2007). Arctic sea ice decline: faster than forecast. *Geophysical Research Letters*, 34, L09501, doi:10.1029/2007GL029703.

3.2: Understand terrestrial ice processes, ecosystem processes, ecosystem services, and climate feedbacks in the Arctic

Lead Author:

Shella Biallas, DOI

Agencies Partners:

DOE, DOI, NASA, NOAA, NSF, SI

Studying diminishing land ice, terrestrial ecosystem processes, ecosystem services, and climate feedbacks will contribute to a better understanding of the cumulative impacts of changes taking place in the Arctic and facilitate more informed decisions in the face of those changes.

Mass drainage of Arctic land ice from the Greenland ice sheet and glaciers and ice caps is a new and poorly understood problem with global implications (Milne *et al.* 2009). Climate feedbacks from ongoing ecosystem changes are only beginning to be understood but are expected to lead to further changes in the global climate system while affecting local communities' abilities to adapt to altered conditions on the ground. The 2004 Arctic Climate Impact Assessment (ACIA 2005) outlined a number of high-priority science needs, including a better understanding of causes of increased Arctic land ice mass loss, changes likely to occur in terrestrial ecosystems, and how those changes will impact local communities. Currently, there are strong research partnerships among U.S. agencies operating in the Arctic. The Landscape Conservation Cooperatives (http://www.doi.gov/lcc/index.cfm), the North Slope Science Initiative (http://www.northslope.org/), and the Alaska Climate Science Center (http://www.doi.gov/csc/alaska/index.cfm), for example, are working to implement an applied-science program to inform management of Arctic natural resources.

The following five research areas will advance existing research participation, enhance coordinated approaches to understanding loss of land ice, climate feedbacks, terrestrial ecosystem processes and services, and support actions to address those changes.

3.2.1 Perform glacial-process studies targeting specific dynamic regimes

Why do this?

Mass loss from the Greenland ice sheet and Arctic glaciers and ice caps, as well as from the Antarctic Ice Sheet, has increased rapidly since the mid-1990s. The combined loss from polar land-ice now accounts for one-third to one-half of sea level rise (Cazenave and Llovel 2010; Church *et al.* 2011; Rignot *et al.* 2011; Jacobs *et al.* 2012). That loss is documented using space-based remote sensing—including altimetry, gravimetry, and Interferometric Synthetic Aperature Radar—as well as field and aircraft studies. Geodetic measurements of continental uplift and Earth rotation support these observations (e.g., Jiang *et al.* 2010; Nerem and Wahr 2011; Mitrovica and Wahr 2011).

The importance of research into the causes of increased Arctic land ice mass loss has been highlighted in numerous recent reports (e.g., http://www.acia.uaf.edu/, http://www.climatescience.gov/Library/sap/sap1-2/final-report/default.htm, http://www.arctic.noaa.gov/reportcard). Loss of Arctic land-ice will result in potentially serious "downstream effects" including regional and global sea-level change (e.g., vulnerabilities of coastal populations, economic and national security infrastructure, and habitats). Understanding the causes of increased Arctic land ice mass loss, their connection to ocean and atmospheric variability, and the implications for current and future sea level will require process studies (this research area), observational studies (3.4), and modeling (3.5).

In the Arctic, the ice mass loss is due to increased surface melting and the acceleration, retreat, and thinning of tidewater glaciers in Greenland (van den Broeke *et al.* 2009) and Arctic tidewater glaciers (Arendt 2011). Though more research is needed to fully understand the complex drivers behind the observed acceleration of tidewater glaciers (Moon *et al.* 2012), evidence suggests that climate change is a major factor. Evidence points to the marine margins of these glaciers as the region from which changes have propagated inland, but the drivers and mechanisms behind the acceleration are still unclear (Vieli and Nick 2011). The significance of this dynamic response has only recently been appreciated and is only beginning to be represented in current-generation ice sheet models. In the 2007 Intergovernmental Panel on Climate Change Fourth Assessment Report (IPCC AR4) report, this shortcoming was identified as the largest source of uncertainty in sea level change projections (Lemke *et al.* 2007). Current projection estimates vary by more than an order of magnitude (Pfeffer *et al.* 2008; Price *et al.* 2011). In addition, the proximity of Greenland to the North Atlantic's dense water formation regions implies that an increasing discharge of freshwater from Greenland can potentially impact the large-scale overturning circulation of the North Atlantic (Dickson *et al.* 2008), a major player in the global oceanic heat transport, with far-reaching climatic implications.

The goals of the process studies are to identify and understand the relevant processes and develop/improve parameterizations for those processes that cannot be resolved in models. Six classe processes are identified, targeting Greenland outlet glaciers and Arctic tidewater glaciers:

1. Ice/ocean boundary layer and plume dynamics: Key measurements and modeling of the turbulent processes and their controls are needed to estimate submarine melt rates and develop appropriate melt rate parameterizations;

2. Fjord circulation and exchanges with the continental shelf: Integrated observational, modeling, and data analysis efforts are needed to understand how the fjord and shelf dynamics impact the ice/ocean boundary and the properties of sea ice, icebergs, and the ice mélange;

3. Glacial hydrology: Knowledge of the supraglacial, englacial, and subglacial hydrology, including discharge of freshwater into the fjord environment, is key to our understanding of ice flow, submarine melt rate, and plume dynamics;

4. Glacier dynamics: Process studies need to address the transition in ice flow from large catchment basins to narrow outlet or tidewater glaciers, in order to understand how the changes in stress-distribution and large-scale bed geometry influence the flow of ice and its supply to the terminus;

5. Calving: Calving plays a crucial role in both ice loss at the terminus and (indirectly) on the acceleration of inland ice flow, but its description remains elusive; and

6. Mass change.

Timeframe:

Mid-term (3-5 years)

Expected Outcomes:

- Improved understanding of basic principles and mechanisms controlling the identified physical processes.

- Direct observations of the meltwater plume at the ice/ocean interface, its spatio-temporal variability, accompanied by plume-resolving simulations.

- Observations of spatio-temporal variability of glacial hydrology (especially supraglacial lakes) and its relation to meteorological conditions, glacier flow, and meltwater plume variability.

- Developed and tested empirical relationship constrained by observations among submarine melt rate, glacier flow, fjord geometry, and circulation and ocean/atmosphere/sea-ice variability.

- Developed and tested empirical relationship constrained by observations between calving and environmental conditions as well as glacier internal stress balances.

- Based on observed relationships, formulate physically based parameterizations of unresolved processes for use in large-scale Earth system models (see 3.4).

- New methods and instrumentation systems, capable of monitoring subsurface melting, calving, subglacial discharge, and ice and sea-water properties in harsh environments.

- Enhance field campaigns with collection of paleo-proxies that provide evidence for past glacier variability.

Milestones:

- Define observational requirements (essential variables, spatio-temporal sampling) for each of the components (ocean/fjord/ice-ocean interface/glacier/atmosphere) for process studies and sustained monitoring (NASA, NOAA; FY2014).

- Initiate 3-5 year interdisciplinary and interagency research initiatives that draw together observational, process modeling, and Earth system modeling expertise to accelerate progress on specific process parameterizations in U.S. Earth system models (NASA, NOAA, NSF; FY2014).

- Report strategies to feed results of process studies into sustained monitoring systems and Earth system models (DOE, NASA, NOAA, NSF; FY2014).

- Foster a U.S. and international research focus on ice/ocean interactions, especially with regards to the Greenland Ice Sheet. (NASA, NOAA, NSF; FY2015).

Science and Technology Gaps and Needs:

- Development of instrumentation systems capable of functioning in harsh glacier terminus and fjord environments capable of withstanding the presence of sea-ice and iceberg, and observational techniques yielding robust measurements with high level of signal-to-noise ratio.

- Improvement of remote sensing technologies and underlying retrieval algorithms.

- Establishment of comprehensive, well-structured, and sophisticated databases and data formats to allow easy access and optimal use of data.

- Development of synergetic approaches (observational, theoretical, modeling, as well as multi-disciplinary) in order to pool resources and make progress in the directions outlined above.

3.2.2 Coordinate and integrate terrestrial ecosystem research efforts

Why do this?

Research on terrestrial ecosystems in the Arctic would greatly benefit from access to the range of existing data systems, information portals, and efforts to identify gaps in dissemination of research data and information. Resource and information sharing among agencies could be significantly enhanced by establishing formal agreements, such as the Memorandum of Understanding between NOAA and USGS (MoU 2011).

Existing sources of information and portals that provide a basis for comprehensive information sharing and coordination include:

- NSF Cooperative Arctic Data and Information Service (CADIS) and Advanced CADIS system (A-CADIS).

- Alaska Data Integration Working Group (ADIWG), formed to examine and address the technical barriers to efficiently integrate and share data within and among participating organizations.

- Polar Data Catalog maintained by the Canadian Cryospheric Information Network (CCIN).

- The North Slope Science Initiative/University of Alaska Geographic Information Network (NSSI/GINA).

- USGS Science Portal.

- The Alaska Ocean Observing System (AOOS).

- The Arctic Council's Conservation of Arctic Flora and Fauna (CAFF) Circumpolar Biodiversity Monitoring Program (CBMP) is developing a distributed and decentralized geo-referenced, web-based portal for communications, data archiving, and information exchange.

- Arctic Portal established as part of the International Polar Year (IPY) activities and currently maintained by Iceland and various Arctic Council Working Groups.

- Arctic, Western Alaska, and Aleutians & Bering Sea Landscape Conservation Cooperatives (LCCs).

- Individual IARPC agencies (DOE, NASA, NPS[8], NWS[9], USGS) maintain substantial databases that inform cross-agency efforts.

Timeframe:

Near-term (1-3 years)

Expected Outcomes:

- Federal and Alaska state agency scientists identify the five most-pressing needs for scientific research in the terrestrial Arctic, in cooperation with Arctic Council efforts.

- Existing information sharing mechanisms, such as data.gov, AON/CADIS, or the NSSI/GINA data catalog, are fully used by scientists in planning and implementing terrestrial-ecosystem research in the Arctic to coordinate with other ongoing and planned efforts, and to expeditiously report their research findings.

- A single information delivery data hub is identified as the primary access point for various portals and databases of U.S. and international Arctic research activities to provide a point of access and collaboration for research activities.

Milestones:

- Databases currently maintained by the Arctic LCC, NSF, NSSI, USGS, and non-governmental groups are made available on one website or via one access point (DOI, NSF; FY2013).

- Review completed and ongoing Arctic studies involving cross-disciplinary collaboration at the principle investigator level and identify any outstanding gaps and needs (NSF; FY2014).

Science and Technology Gaps and Needs:

Federal agencies should agree upon and adopt a single access point to existing and future information-sharing databases and portals and share them with cooperating state, local, and international partners.

8. National Park Service (Department of Interior)
9. National Weather Service (NOAA, Department of Commerce)

Stored long-term observing data and associated information should be accessible to and easily managed by contributing partners.

3.2.3 Identify and study key sites where climate feedbacks are active, including permafrost, snow, hydrates, and glaciers

Why do this?

The Arctic cryosphere regulates local, regional, and global climate and provides vital ecosystem services to communities. It also stores vast quantities of carbon (including in the form of methane) that may be released as temperatures increase and as ecosystem processes—such as wildland fires—change. A warming climate is having significant impacts on the Arctic cryosphere, such as decreased snow cover and thawing permafrost in the terrestrial environment. While terrestrial feedbacks in some locations have been studied, climate feedbacks in the region are generally still poorly understood (Arctic Monitoring and Assessment Program 2011).

The 2011 report, *Snow, Water, Ice, and Permafrost in the Arctic* (SWIPA), from the Arctic Monitoring and Assessment Program (AMAP) of the Arctic Council identified a number of areas where additional U.S.-supported research could improve understanding of the cryosphere (http://amap.no/swipa/). For example, better quantifying the intensity of cryosphere-climate feedbacks would lead to better predictions of the degree and rate of changes in the cryosphere and Arctic environment. It will also be critical to understand how Arctic communities will be affected by cryospheric changes, including those that influence migration patterns of subsistence species, utility of areas of traditional use, and the stability of infrastructure such as roads, airports, pipelines, and buildings. Other research is needed to understand how climate-driven changes in different parts of the cryosphere are affecting Arctic ecosystems—a critical first-step toward being able to predict how particular species (including those important for subsistence) will respond to new conditions.

Particular sites where climate feedbacks are observable can be identified by enhancing existing networks that build capacity for identifying, understanding, predicting, and responding to diverse environmental changes throughout the Arctic. For example, the Arctic LCC is identifying a suite of biophysical process topics most relevant to forecasting species and habitat response, which will guide recommendations for priority monitoring and modeling products. The Arctic LCC will recommend candidate sites for inclusion in a terrestrial ecosystem-monitoring network for northern Alaska that addresses the link-ages between physical drivers and biological responses. Implementing those recommendations will require the combined efforts and support of multiple agencies. The World Climate Research Program's Climate and Cryosphere (CliC) Project supports and coordinates international research on the interaction between the cryosphere and the rest of the climate system, focusing on identifying patterns and rates of change (http://www.climate-cryosphere.org/en/). Also, the International Network for Terrestrial Research and Monitoring in the Arctic (INTERACT) studies 32 sites in 14 countries, including all Arctic nations, with a goal of building capacity to improve understanding of environmental changes in the Arctic (http://www.eu-interact.org/).

Timeframe:

Mid-term (3-5 years)

Expected Outcomes:

- Cross-agency agreement on geographical areas in the U.S. Arctic where additional research on climate feedbacks and impacts on the cryosphere is most needed.

- Increased understanding of geophysical and ecosystem responses to changing climate and cryosphere change, which informs management decisions and subsistence uses.

Milestones:

- Consult with local communities on the geographical areas of traditional use that are most impacted by changes in the terrestrial cryosphere, as well as the types of changes—permafrost thaw, coastal erosion, vegetation—most relevant to local communities (DOI; FY2013).

- Identify and compile Geographic Information Systems (GIS) data layers of existing climate feedback research in the Arctic and compare with areas of important traditional use (DOI; FY2012).

- Complete Wildlife Potential Habitat Forecasting Framework (WILDCAST) projections of potential land cover and habitat changes in Northern Alaska (USGS, FY2012).

- Assess existing tools and methods for measuring and mapping the effects of cryosphere changes on Arctic ecosystems and communities, including water levels, sea level rise, salt water intrusion, and coastal inundation (DOI, NSF; FY2013).

- Complete improved coastal map and high precision Digital Elevation Models (DEM) for western and northern Alaska in order to better understand coastal erosion, storm surges, and sea level rise. (DOI, NOAA, USCG; FY2013).

Science and Technology Gaps and Needs:

Traditional knowledge can help detect changes in terrestrial Arctic ecosystems and guide adaptation. As the cryosphere changes, climate feedbacks become more complex, changes in Arctic systems will occur more rapidly, and integrating traditional knowledge into decisions will be as important as ever. Other needs include systematically collecting information about the effects of cryospheric change on human society and greater engagement between the scientific community and Arctic residents.

Improved networks are necessary to observe many of the long-term environmental changes taking place in the Arctic, including those pertaining to climate feedbacks (see section 3.4). Coordinating measures of changes in ecosystems and their components (e.g., species of concern) from *in-situ* and discrete observations with wider-ranging environmental parameters from remote instruments remains a challenge. Such coordination will improve the ability to relate geophysical change patterns to processes affecting ecosystems occurring across a range of spatial scales.

3.2.4 Investigate the frequency and severity of wildland fires in the Arctic

Why do this?

Climate changes are affecting the frequency, extent, and severity of Arctic wildland fires and will have cascading effects on other ecosystem processes (Olsen *et al.* 2011). Wildfire is the primary disturbance in boreal forests of interior Alaska and is also a disturbance in the tundra regions of the U.S. Arctic. Fire frequency and severity are primary determinants of vegetative succession trajectories and, subsequently, the rates of carbon sequestration and loss in boreal ecosystems. Fire frequency and severity are likely to increase in tundra ecosystems as plant biomass and productivity increase with the lengthening growing season. Vegetation succession following fire in tundra ecosystems, however, is not well understood (Final Report of the Joint Fire Study Program 2011). In general, there is evidence that observed increase in tall shrub cover (Tape *et al.* 2006) is linked to an increase in fire frequency (Higuera *et al.* 2008). Fire regimes and impacts have been observed on fine spatial scales at point locations in Alaska, but the collective impacts of wildland fire on vegetation, carbon, wildlife, air quality, permafrost degradation, and biogeochemical cycles across the landscape are poorly understood, yet critical to regional fire management strategies.

The 2007 Anaktuvuk River fire was the largest tundra fire to take place in the United States since recording began in 1950 and had major implications both for local ecosystems and the global carbon system. Several agencies have funded investigations of the severity of and ecological response to the Anaktuvuk River fire. That type of research, however, has not been a priority for resource managers. A recent study found the Anaktuvuk fire released carbon into the atmosphere about 100 times faster than it usually escapes from the ground in the Arctic summer, and released more than two million tons of CO_2 (Mack 2011). The fire was deep and severe for a tundra burn (Mack *et al.* 2011), and lichen cover was drastically reduced, thereby reducing winter forage for caribou. Because caribou foraging on winter range tend to avoid burned areas for up to 50 years (Joly 2007), large-scale burns may change seasonal distribution of the animals, thereby changing their availability to subsistence users.

Timeframe:

Mid-term (3-5 years)

Expected Outcomes:

- An integrated understanding of the trends and impacts of Arctic wildland fires to inform policy and land-management decisions.

Milestones:

- Identify and inventory existing scientific research on Arctic wildland fires (DOI; FY2012).

- Consult with local communities and indigenous groups on science needs pertaining to Arctic wildland fires and their impacts on cultural and subsistence needs (DOI; FY2012).

- Develop strategies/projects to identify succession stages of tundra communities following a wildfire (DOI, NSF; FY2013).

- Ensure coordination in the development of models that incorporate feedback from fire models with models of surface vegetation and organic layer properties and permafrost and soil conditions, incorporating hydrologic information as appropriate (DOE, DOI, NSF; FY2016).

Science and Technology Gaps and Needs:

While the 2007 Anaktuvuk River fire was an event of unprecedented scale on the North Slope, it is not clear whether ongoing climate changes will lead to additional severe fires. Paleoecological work assessing fire-deposited charcoal in lake cores is underway and should contribute to improved under-standing of fire regime in that area. As fire-return intervals have historically been several hundreds of years, detecting less than very dramatic changes in long-term fire regime, or fire severity, may not be possible. Credible prediction of the potential effects of changes in fire regime, or fire management policy, on caribou distribution and subsistence use will likewise be challenging. The Alaska Integrated Ecosystem Modeling project, through the Alaska Climate Science Center and the Western, Arctic, and Northwest Interior Forest LCCs, is creating a framework to link existing models to produce a single integrated platform that simulates vegetation succession, disturbance regimes, hydrology, and permafrost dynamics. With broad support, this work can contribute to state-wide maps that forecast ecological conditions under specified climate scenarios. Data also are needed to improve response to fires and understanding of fire behavior include greater coverage in lightning-strike detection, particularly in the central Arctic, and enhanced distribution of weather-station observation sites that collect temperature and precipitation data.

3.2.5 Conduct socio-economic research to understand ecosystem services as increased warming changes the Arctic tundra

Why do this?

More than 86% of communities in rural Alaska have been affected by serious erosion and flooding (GAO 2003). Forced relocation of some communities comes at a high cost, and sites must be selected that will remain stable indefinitely. Local residents are being forced to address immediate needs, develop adaptation strategies for the future, and also balance social, cultural, and economic demands and expectations simultaneously (see section 3.6).

Additional environmental stressors brought on by increased human activity in the area are also affecting Arctic communities. Such rapid change affects supply of subsistence foods and storage capability, integrity of local infrastructure, and social and cultural systems (http://www.arctichealth.org/ccNorthernCommunities). At the same time, economic and political pressures to fulfill global energy needs place a burden on Arctic residents.

More research is needed on the social and economic impacts of ongoing environmental changes and their implications for local communities. Such information will provide the framework and data needed to develop proactive plans for protecting, managing, and adapting to a fragile and changing Arctic environment.

Timeframe:

Mid-term (3-5 years)

Expected Outcomes:

- A survey of the state of knowledge on terrestrial ecosystem services in the Arctic including iden-tification of gaps in research on the importance of ecosystem services for Northern communities.

- A set of robust and reliable social indicators are established, and relevant data are collected to predict, monitor, and mitigate the effects of climate change on Alaskan Arctic communities.

Milestones:

- Support the outcomes and recommendations of the Arctic Social Indicators Project (http://www.svs.is/asi/Implementation/Project%20description%20II.htm) within the United States.

  - Develop a meta-database, published electronically, that identifies baseline Arctic social indicators already monitored by national agencies (DOI, NSF, SI; FY2013).

  - Establish an international task force of Arctic researchers to ensure social data are collected in a way that allows statistically valid comparisons among Arctic communities (DOI, DOS, NSF, SI; FY2012).

  - Collect indigenous and non-indigenous regional data on the socio-economic implications of climate change at 5-year intervals (DOI, NIH, NSF, SI; FY2013).

Science and Technology Gaps and Needs:

Socio-economic data at the village and regional level are lacking as are direct investigations of the links between cause and social effects. The mechanisms of cause and effect are often complex and multi-factoral. Small population size at the village and region level may preclude acquisition of statistically significant data for some indices.

References

ACIA (2005). Arctic Climate Impact Assessment: Scientific Report. 1042 pp. Cambridge University Press, UK.

Arctic Health. University of Alaska—Anchorage. Compilation of Alaska Northern Communities Resilience and Adaptation Research—http://www.arctichealth.org/ccNorthernCommunities.

Arctic Monitoring and Assessment Program. (2007). Oil and Gas in the Arctic: Effects and Potential Effects Volume 1 (Chapter 2—Social and Economic Effects).

Arctic Monitoring and Assessment Program. (2011). Working chapter 11.1 of Arctic Council SWIPA report.

Arendt, A. A. (2011). Assessing the status of Alaska's glaciers. *Science*, 332, 1044-1045. doi: 10.1126/science.1204400.

Cazenave, A., & Llovel, W. (2010). Contemporary sea level rise. *Annu. Rev. Mar. Sci.*, 2, 145-173. doi:10.1146/annurev-marine-120308-081105.

Church, J. A., *et al.* (2011). Revisiting Earth's sea-level and energy budgets from 1961 to 2008. *Geophys Res Lett.*, 38, L18601.

Dickson, R. R., Meincke, J., & Rhines, P. (2008). Arctic-Subarctic Ocean Fluxes. Springer.

Government Accountability Office Report. (2003). ALASKA NATIVE VILLAGES: *Most Are Affected by Flooding and Erosion, but Few Qualify for Federal Assistance.* http://www.gao.gov/new.items/d04142.pdf. Last accessed Jan. 22, 2013.

Higuera, P.E., *et al.* (2008). Frequent Fires in Ancient Shrub Tundra: Implications of Paleorecords for Arctic Environmental Change. *PLoS ONE* 3(3), e0001744. doi:10.1371/journal.pone.0001744.

Jacob, T., Wahr, J., Pfeffer, W. T., & Swenson, S. (2012). Recent contributions of glaciers and ice caps to sea level rise. *Nature*, 482, 514-518. doi:10.1038/nature10847.

Jiang, Y., Dixon, T.H., & Wdowinski, S. (2010). Accelerating uplift in the North Atlantic region as an indicator of ice loss. *Nature Geosci.*, 3, 404-407. doi:10.1038/ngeo845.

Joly, K., Bente, P., & Dau, J. (2006). Response of Overwintering Caribou to Burned Habitat in Northwest Alaska. *Arctic.* 60(4) (December 2007) 401–410.

Lemke, P., Ren, J., Alley, R. B., Allison, I., Carrasco, J., Flato, G., Fujii, Y., Kaser, G., Mote, P., Thomas, R.H., & Zhang, T. (2007). Observations: Changes in Snow, Ice and Frozen Ground. In: Climate Change 2007: The Physical Science Basis. Contribution of Working Group I to the Fourth Assessment Report of the Intergovernmental Panel on Climate Change. [Solomon, S., D. Qin, M. Manning, Z. Chen, M. Marquis, K.B. Averyt, M. Tignor and H.L. Miller (eds.)]. Cambridge, UK. & New York, NY: Cambridge University Press.

Mack, M. C., Bret-Harte, M. S., Hollingsworth, T. N., Jandt, R. R., Schuur, E. A. G., Shaver, G. R., & Verbyla, D. L. (2011). Carbon loss from an unprecedented Arctic tundra wildfire. *Nature* 475, 489-492.

Memorandum of Understanding between the USGS and NOAA to Coordinate and Cooperate in Activities Involving Physical and Biological Sciences and Environmental Studies. Signed Jan. 9, 2011.

Milne, G. A., Gehrels, W. R., Hughes, W., & Tamisiea, M. E. (2009). Identifying the causes of sea-level change. *Nature Geosci.*, 2, 471-478.

Mitrovica, J. X., & Wahr, J. (2011). *Ice Age Earth Rotation.* Annu. Rev. Earth Planet. Sci., 39, 577–616. doi:10.1146/annurev-earth-040610-133404.

Mix, A. C., Samelson, R., & Padman, L. (Eds.). (2012). Interdisciplinary Approaches to Understanding Atmosphere/Ocean/Ice Shelf/Ice Sheet Interactions. A Workshop Report to NSF.

Moon, T., Joughin, I., Smith, B. & Howat, I. (2012). 21st-Century evolution of Greenland outlet glacier velocities. *Science*, 336, 576-578.

Nerem, R. S., & Wahr, J. (2011). Recent changes in the Earth's oblateness driven by Greenland and Antarctic ice mass loss. *Geophys. Res. Lett.*, 38, L13501. doi:10.1029/2011GL047879.

Olson, D., *et al.* (2011). Final Report to the Joint Fire Science Program: Compiling, Synthesizing and Analyzing Existing Boreal Forest Fire History Data in Alaska. July 2011. http://www.firescience.gov/projects/06-3-1-26/project/06-3-1-26_final_report.pdf. Last accessed Jan. 22, 2013.

Pfeffer, W. T., Harper, J. T., & O'Neel, S. (2008). Kinematic constraints on glacier contributions to 21st-century sea-level rise. *Science*, 321, 1340-1343.

Price, S. F., Payne, A. J., Howat, I. M., & Smith, B. E. (2011). Committed sea-level rise for the next century from Greenland ice sheet dynamics during the past decade. *P. Natl. Acad. Sci. USA*, 108, 8978–8983.

Rignot, E., Velicogna, I., van den Broeke, M. R., Monaghan, A., & Lenaerts, J. (2011). Acceleration of the contribution of the Greenland and Antarctic ice sheets to sea level rise. *Geophys. Res. Lett.*, 38, L05503. doi:10.1029/2011GL046583.

Tape, K., Sturm, M., & Racine, C. (2006). The evidence for shrub expansion in Northern Alaska and the Pan-Arctic. *Global Change Biology*, 12, 686-702.

The Arctic Social Indicators Project. http://www.svs.is/asi/Implementation/Project%20description%20II.htm van den Broeke, M., *et al.* (2009). Partitioning recent Greenland mass loss. *Science*, 326, 984. doi: 10.1126/science.1178176.

Vieli, A., & Nick, F. M. (2011). Understanding and modeling rapid dynamical changes of tidewater outlet glaciers: issues and implications. *Surv. Geophys.* 32(4-5), 437–458. doi:10.1007/s10712-011-9132-4.

3.3 Understand atmospheric surface heat, energy, and mass balances

Lead Authors:
Richard Eckman, NASA
Taneil Uttal, NOAA
Sandy Starkweather, NOAA

Agency Partners:
DOE, NASA, NOAA, NSF

Year-to-year and longer-scale trends and variability in surface-air temperature tend to be larger in the Arctic than in other parts of the globe. This Arctic amplification phenomenon is recognized as an inherent characteristic of the global climate system (Serreze and Barry 2011); the causes are believed to include complex interactions associated with heat exchange between the atmosphere and ocean (with its changing sea-ice extent), meridional heat transport, and radiative forcing from atmospheric constituents. Arctic atmospheric processes are influenced by unique features (polar night, high albedo surfaces, and atmospheric stability) that can change the sign and magnitude of aerosol and cloud radiative forcing relative to low latitudes. Important uncertainties in the sign and magnitude of these forcings (e.g., IPCC AR4) provide the rationale for coordinating and improving our integrated understanding of Arctic atmospheric processes. This plan addresses those uncertainties through milestones aimed at improving process representations in models, reducing uncertainty in model outputs, and developing long-term observational data sets.

Remote sensing and *in-situ* observations at long-term observatories complement each other and contribute to documenting and understanding long-term trends. NASA and NOAA remote sensing assets offer an unique resource for observing Arctic atmospheric composition and radiative forcing. DOE, NSF, and NOAA support key ground-based observatories providing long-term data sets and campaign studies that are critical resources for addressing key climate model uncertainties such as clouds, aerosols, and impacts of short-lived climate forcers (SLCFs). These land-based observatories are also valuable resources for validating satellite observations. Data from these observatories also are available through the International Arctic System for Observing the Atmosphere (IASOA). During the International Polar Year, IASOA developed a major new international observatory facility in Tiksi, Russia, that significantly extends circumpolar coverage. DOE, NASA, NSF, and NOAA field and aircraft campaigns have contributed to an integrated understanding of the Arctic atmosphere and improved model parameterizations and the value of remote-sensing data products. DOE- and NSF-supported predictive modeling contributes to long-term understanding of regional and global sensitivities to aerosol loading and cloud processes, have revealed the relative contributions of short-lived climate forcers (discussed in more detail in the next section), and have helped to develop an integrated picture of the atmospheric interactions with terrestrial, oceanic, and cryospheric systems. Coordinated approaches around the Arctic, particularly those supported by international partnerships, are critical to developing a regionally coherent understanding of both how and why the Arctic atmosphere is changing. Jointly, these activities are improving our understanding of the unique role of the Arctic atmosphere in influencing the Arctic surface system as well as regional and global weather and climate.

3.3.1 Improve understanding of short-lived climate forcers (SLCFs); source regions, direct and indirect effects, and net impact on Arctic warming

Why do this?

Short-lived climate forcers (SLCFs) including black carbon, methane, and ozone are atmospheric constituents with relatively short residence times (days to years) and that warm the climate. These forcers are thought to have an enhanced influence on Arctic radiative forcing relative to mid-latitudes (Quinn *et al.* 2008). For example, black carbon can change Arctic radiation balances by either direct forcing by the radiative profile of the atmosphere, by aerosol-cloud indirect effects, or by lowering the albedo of (typically) bright Arctic surfaces after deposition or, potentially, by hastening the thaw of snow and ice. Investigating the relative contributions of these three mechanisms and the effects of black carbon requires additional attention. The Arctic contains vast amounts of methane locked in permafrost deposits and marine hydrates, with an uncertain potential for release into the atmosphere. As a greenhouse gas, methane is approximately twenty times more effective at trapping heat than is carbon dioxide over a 100-year period. Understanding current methane emissions and potential scenarios under a warmer Arctic is imperative. Many global circulation models do not take into account carbon feedback loops from Arctic tundra, where warming causes carbon release from thawing and decaying tundra that, in turn, could further accelerate carbon release. Ozone is both an air pollutant that impacts human health as well as a greenhouse gas. Both Arctic and remote sources produce ozone and its precursors. Boreal forest fires and increasing human activity will increase ozone precursors in the region. Further research is required to understand sources of ozone precursors as well as the oxidation capacity of the Arctic atmosphere. Ozone depletion events have been related to the deposition of mercury, which is a significant and toxic pollutant.

Existing agency capabilities to study SLCF's include satellite instruments that monitor the long-range transport of mid-latitude pollution to the Arctic, in addition to detecting fires and their smoke plumes. These instruments measure aerosol optical depths, other aerosol properties, and collocated cloud properties. The Moderate Resolution Imaging Spectroradiometer (MODIS) and the Multi-angle Imaging SpectroRadiometer (MISR) provide an aerosol record dating from 2000, while the Atmospheric Infrared

Sensor (AIRS) provides a carbon monoxide record from 2002. The Cloud-Aerosol LIDAR and Infrared Pathfinder Satellite Observation (CALIPSO) satellite Cloud-Aerosol LIDAR with Orthogonal Polarization (CALIOP) instrument provides high-resolution vertical profiles of aerosols and clouds from 2006. Flying in formation with Aqua (MODIS, AIRS), Aura and GCOM-W1, the multi-sensor "A-Train" provides near simultaneous measurements of a variety of parameters. IASOA network observatories include proxy and direct measurements of black carbon, as well as direct measurements of methane and ozone at limited locations. About a dozen Aerosol Robotic Network (AERONET) and AeroCan (AERosol CANada) sunphotometer sites located at latitudes poleward of 60°N monitor aerosol amount and type in seasons when the sun is above the horizon. The DOE Next Generation Ecosystem Experiment (NGEE) is focusing on measurements for improving the model simulations of climate-related emissions of SLCFs. DOE, NASA, and NOAA aircraft campaigns have also contributed to an integrated understanding of the Arctic atmosphere. The NASA Arctic Research on the Composition of the Troposphere from Aircraft and Satellites (ARCTAS), for example, provided new characterization of bidirectional reflectance distribution functions for aerosols over Arctic surfaces, validated CALIOP LIDAR aerosol sensing for a range of conditions, and provided detailed characterization of the optical properties of aerosols from boreal fires.

Timeframe:

Mid-term (3-5 years)

Expected Outcomes:

- Sustained and improved satellite-observation capabilities focused on SLCF's (including black carbon, methane, and ozone).

- Enhanced *in-situ*, long-term observations of SLCF's including a methane-observing network.

- Improved modeling of SLCF transport and lifetime.

Milestones:

- Support process studies and campaigns to validate current satellite measurements, such as aerosol products from MODIS, MISR, and Visible Infrared Imager Radiometer Suite (VIIRS) and methane and near-surface ozone from the Tropospheric Emission Spectrometer (TES) on Aura (DOE, NASA, NOAA; FY2016).

- Develop pan-Arctic synthesis of SLCF's from current observations focused on concentrations, sources, and radiative impacts (NOAA, NSF; FY2014).

- Develop needs assessment for improved transport-modeling capability (DOE, NSF; FY2014).

- Support SLCF's source identification through transport and regional modeling using satellite and suborbital data to constrain the models (DOE, NSF; FY2017).

- Support black carbon source identification through chemical composition measurement at key observatory locations and aerosol mapping from space with MODIS and MISR (DOE, NASA, NOAA; FY2017).

- Support black carbon radiative impact studies through *in-situ* measurements at key observatory locations and modeling of light scattering, absorption, and aerosol optical depth (DOE, NOAA; FY2017).

- Develop needs assessment for an Arctic methane-observation network (DOE, NASA, NSF; FY2013).

- Increase spatial density of Arctic methane measurements (DOE, NOAA, NSF; FY2017).

Science and Technology Gaps and Needs:

NASA operates a variety of sensors that monitor trends and transport of SLCF's. Satellite retrievals of SLCF's offer poor resolution near the Earth's surface and must be integrated with higher-resolution *in-situ* measurements. Agencies should support field validation campaigns to obtain direct information about aerosol and surface properties as well as other activities that reduce uncertainty in retrievals by satellites.

Large spatial gaps exist in the current network of *in-situ* measurements for SLCF's. Ground-based observatories could make stronger contributions to understanding the role of black carbon by building capacity to monitor the mass concentrations and chemical composition of aerosols in addition to current light-absorbing proxy measurements. Locations should be based on sampling representative air-mass trajectories. Given the spatial extent and diversity of methane sources, a much broader network of sustained methane measurements is required to monitor long-term methane trends, understand the processes that emit methane, and better constrain inverse model studies of the methane budget.

3.3.2 Improve understanding of processes that control the formation, longevity, and physical properties of Arctic clouds, including the effects of—and sensitivities to—aerosols

Why do this?

Unlike other atmospheric features, clouds and aerosols are unevenly distributed in space and participate in highly integrated processes. Clouds, particularly those at low levels, occur frequently throughout the Arctic. They are particularly susceptible to aerosol influences on both liquid-droplet and ice-crystal nucleation, which impacts cloud formation, persistence, physical properties, and precipitation. Substantial uncertainty surrounds which modes of nucleation are operating under the varying conditions in the Arctic and how they are linked to aerosol composition and sources. Due to the prevalence of low sun angles and high surface albedos, Arctic aerosol and cloud radiative forcings typically have different signs than at lower latitudes. Specifically, they tend to warm the surface predominantly by trapping long-wave radiation more efficiently than they cool the surface by reflecting sunlight. Aerosols can change cloud cover, thickness, and brightness. These perturbations, thus, have direct implications for the net radiative balance at the Earth's surface and top of the atmosphere. Arctic long-wave radiative forcing from low-lying clouds has been identified as an important controller of onset and duration of surface melt impacting the mass budgets of sea ice and ice sheets and the seasonal extent of snow cover. Cloud precipitation processes are a fundamental component of the Arctic hydrological cycle. Due to numerous, poorly known, complex processes involving clouds and aerosols, their misrepresensation in models lead to large uncertainties in climate simulations of (i.e., IPCC AR4). To improve these modeling deficiencies, it is imperative to characterize and understand basic cloud and aerosol properties and their interactions within the system.

U.S. agencies (DOE, NASA, NOAA, and NSF) employ satellite, ground-based, and *in-situ* assets to observe Arctic clouds and aerosols. Passive measurements from space are challenging, particularly during winter, and over snow and ice surfaces, due to low optical and thermal contrast among clouds, aerosols, and the underlying surface. Using satellites to assess the impact of aerosols on Arctic clouds is an even greater challenge for passive remote-sensing instruments due to persistent cloud cover, poor vertical discrimination, bright surfaces, relatively low aerosol column abundances, and low solar-illumination angles, but these measurement limitations are significantly improved by combining passive instruments, such as MODIS (measuring ambient radiation), with active instruments, such as CALIOP and CloudSat (measuring emitted LIDAR and radar returns). In addition to providing an unparalleled aerosol record going back to 2000, the MODIS and MISR satellite sensors monitor the long-range transport of pollution and smoke into the Arctic. Ground-based observations, however, provide a very important dataset of Arctic clouds. These measurements are being used to improve our knowledge of cloud processes, and thereby improve our ability to represent these processes in numerical models. Furthermore, these observations are also providing detailed climatologies, albeit in only a small number of locations, of cloud occurrence and phase as a function of height, as well as cloud particle size and water path.

Important progress has also been made towards understanding cloud-aerosol interactions and properties via aircraft campaigns such as the recent NASA Arctic Research on the Composition of the Troposphere from Aircraft and Satellites (ARCTAS), the DOE Indirect and Semi-Direct Aerosol Campaign (ISDAC), and the NOAA Aerosol, Radiation, and Cloud Processes affecting Arctic Climate (ARCPAC). For example, ARCTAS provided new characterizations useful for energy balance assessment and a better understanding of aerosol radiative effects, which help to improve satellite measurements. ARCPAC and ISDAC both provided important new perspectives on aerosol composition and transport sources with important implications for cloud processes.

The International Arctic System for Observing the Atmosphere (IASOA) is a pan-Arctic consortium of flagship ground-based observatories that are supported by DOE, NOAA, NSF, Environment Canada (EC), the Russian Federal Service for Environmental and Hydrometeorolgical Monitoring (*Roshydromet*), the Finnish Meteorological Institute (FMI), and other government and non-government contributors from Arctic and non-Arctic countries. The observatories operate sophisticated instruments that support sustained, high-resolution, and simultaneous observations of clouds, aerosols, atmospheric structure, and the surface-energy balance. These year-round observatories provide excellent platforms for both long-term and campaign-based process studies. DOE's Atmospheric Radiation Measurement (ARM) Climate Research Facility in Barrow, Alaska, for example, provides a long-term record of highly valuable simultaneous measurements of cloud microphysical and macrophysical parameters, aerosols, and surface radiation. Recent additions of scanning radars, powerful LIDARs, flux measurements, and aerosol instruments further enhance ARM's measurement capabilities over an extended volume near Barrow. Measurements from that site have been the basis for several successful international model inter-comparison projects focused on cloud-resolving and single-column model simulations, whose goals are to improve Arctic cloud microphysics parameterizations and to better understand indirect effects of aerosols. Instrumentation at the new ARM site at Oliktok, Alaska, mirrors the Barrow site with the added capability of instrumented unmanned aerial vehicles (UAVs). The focus of work at the Oliktok site is to examine clouds and aerosols over land, sea, and ice, as well as the coupled atmosphere-cloud-terrestrial Arctic systems. The NSF has established an ARM-like site at Summit Station on the Greenland Ice Sheet. NOAA and EC operate a similar ARM-like site at Eureka, Canada. Ground-based observatories, in general, contribute unique and valuable information for use in model evaluation and development, for validation of satellite observational methods, and for long-term monitoring of Arctic atmospheric properties.

Timeframe:

Mid-term (3-5 years)

Expected Outcomes:

- Evaluation, improvement, and development of cloud and aerosol parameterizations in climate and weather prediction models.

- Sustained and improved satellite-observation capabilities of cloud and aerosol properties, with verification using surface observations.

- Sustained and enhanced ground-based observations including simultaneous measurements of clouds and aerosols.

- Improved understanding of how clouds respond to changing levels of aerosols and sea ice.

- Synthesis of data sets that provide detailed descriptions of clouds and aerosols.

Milestones:

- Support sustained and enhanced ground-based measurements of cloud and aerosol properties (DOE, NASA, NOAA, NSF; FY2017).

- Conduct intensive, short-term, ground-based, and airborne field experiments to quantify the impact of aerosols on clouds, conduct detailed process studies, and provide validation data sets for remote sensing data (DOE, NASA, NOAA, NSF; FY2017).

- Support synthesis activities to develop long-term observational cloud and aerosol data sets from ground-based and satellite platforms to evaluate model parameterizations (DOE, NASA, NOAA, NSF; FY2017).

- Use observational data sets to constrain process-model studies and conduct detailed model inter-comparisons to advance parameterization development (DOE, NSF; FY2017).

- Support laboratory studies to examine cloud-particle nucleation processes (DOE, NSF; FY2014).

- Support observations and modeling activities to improve understanding of transport of aerosols from remote regions to the Arctic (NASA, NSF; FY2017).

Science and Technology Gaps and Needs:

At many scales, models do not represent cloud and aerosol processes well. This general lack is based upon observational data sets of insufficient length, complexity, and spatial representation, as well as insufficient process-resolving models. Specific gaps in understanding include processes related to aerosol sources and transport, basic cloud-aerosol interactions, cloud-phase partitioning, the influence of heat and moisture advection on cloud formation, interactions between clouds and atmospheric structure, and determining the relative contributions of different cloud types to precipitation totals.

To address the need to advance physical parameterizations in numerical climate and weather-prediction models, measurements related to cloud-aerosol-radiation processes at DOE, IASOA, NASA, NOAA, and NSF observatories should be sustained and enhanced. These measurements should be supplemented by intensive campaigns (DOE, NASA, NOAA, and NSF) focused on enhancing ongoing, long-term measurements and targeting specific processes or hypotheses. To expand the spatial footprint of these observations, DOE, NASA, NOAA, and NSF are funding technology development through the Small Business Innovation Research (SBIR) program to enable cloud and aerosol measurements from UAVs and balloon-borne platforms. Enhanced emphasis should also be placed on validation and refinement of satellite measurement technologies for observing aerosol characteristics and transport, as well as characterizing clouds throughout the troposphere (NASA, NOAA). Satellite research would also greatly benefit from passive measurements from the Suomi National Polar-orbiting Partnership (NPP) mission launched in 2011, the forthcoming Joint Polar Satellite System-1 (JPSS-1), and international collaboration with European countries and Japan around future active measurements from the EarthCare mission. Ideally, development of a more advanced satellite-based, multi-angle, multi-spectral, polarimetric imager should be considered. Finally, where appropriate, agencies should support evaluation, utilization, and advancement of smaller-scale models that can be used in both process studies and as intermediaries for parameterization development.

3.3.3 Develop an integrated understanding of Arctic atmospheric processes, their impact on the surface energy budget, and their linkages with oceanic, terrestrial, and cryospheric systems

Why do this?

Numerical models are used to understand and predict important processes such as the decline of Arctic sea ice, linkages between Arctic conditions and lower-latitude weather, and the general amplification of climate change in the Arctic. Developing models of sufficient quality, however, relies on building a system-level understanding of Arctic climate that includes detailed knowledge about the Arctic atmosphere and surface and their many interacting processes. Current models face considerable difficulties when representing Arctic atmospheric processes related to boundary layer structure, cloud formation, and aerosol-cloud interactions—all of which interact critically with the surface. The general inabilities of models to accurately represent the observed decline in Arctic sea ice and to properly capture its causes represent one tangible example. Such modeling difficulties result directly from limited, system-level observations and could be remedied by comprehensive and coordinated measurements of all contributing components.

Process- and system-level understanding requires coordination and expansion of observational capabilities along with complementary process-modeling activities. Satellite measurements—a cornerstone of observation—provide valuable spatial coverage of many key parameters with linkages to those from lower-latitudes. They are often inadequate, however, for quantifying basic parameters such as surface temperatures and providing the level of detail needed for coordinated radiation, cloud, aerosol, and other atmospheric measurements. Additionally, many polar-orbiting satellites take insufficient measurements in the area within 800 kilometers of the North Pole. Although ground-based observations and aircraft campaigns may be spatially and/or temporally limited, they provide the types of measurements needed to characterize many of the necessary processes in high detail. Most existing knowledge on Arctic atmosphere-surface interactions is biased towards land-based or coastal observations and processes. Thus, the challenge is to coordinate existing and new interagency observational abilities from satellite, aircraft, and the ground to produce the comprehensive, process-level observations of the Arctic system needed to improve numerical models.

Timeframe:

Mid-term (3-5 years)

Expected Outcomes:

- Sustained and improved satellite-observation capabilities focused on atmospheric interactions with the surface and coordination of satellite-measurement capabilities.

- Improved understanding of the two-way relationship between sea ice and clouds.

- Sustained and enhanced ground-based observations emphasizing simultaneous measurements of clouds, aerosols, atmospheric structure, and surface-energy budget in land- and ocean-based environments.

- Improved representation of Arctic systems in climate and weather-prediction models.

Milestones:

- Support model-component development and advancement of fundamental knowledge of the key processes that regulate aerosol and cloud impacts on the atmospheric- and surface-energy budgets (DOE, NASA, NOAA, NSF; FY2015).

- Support research activities that integrate Arctic processes in regional and global models (DOE, NASA, NOAA, NSF; FY2017).

- Coordinate interdisciplinary campaigns to study the Arctic climate system as a whole (DOE, NSF; FY2017).

- Increase use of UAV (unmanned aerial vehicles) platforms for targeted observations of Arctic processes (DOE, NASA, NOAA; FY2015).

Science and Technology Gaps and Needs:

Three prominent gaps exist in research on Arctic atmospheric processes and their interactions with the broader Arctic climate system: 1) a lack of comprehensive observations at specific locations; 2) a general dearth of process-level observations over sea-ice environments; and 3) limited routine observations in the central Arctic Basin.

Barrow, Alaska, is a primary hub for Arctic research in the United States. Benefiting from substantial observatory and campaign efforts by DOE, NASA, NOAA, and NSF, Barrow likely contains the most comprehensive Arctic atmospheric observatory capabilities worldwide and, in many ways, is a model for such observations. This comprehensive level, however, is not currently available at other key Arctic IASOA observatories operated in Canada, Greenland, Russia, and Northern Europe. Efforts should be made to broaden observational capabilities at additional Arctic locations, leveraging existing industrial monitoring sites where appropriate. The new ARM site at Oliktok, Alaska, with instrumentation mirroring the Barrow site—but with the added capability of instrumented UAVs—should attract other agency contributors. Additional instrumentation from other agencies, such as NASA, NOAA, and NSF, would provide more-comprehensive observations of clouds and aerosols over land, sea, and ice, as well as the coupled atmosphere-cloud-terrestrial Arctic systems.

The central Arctic Basin represents a substantial spatial coverage gap that limits models of important processes, such as those related to sea-ice decline. Multi-year, detailed, and comprehensive measurements, extending from the ocean through the sea ice and into the atmosphere, are critically needed to provide a process-level understanding of the complex regional systems of interactions and feedbacks that cannot be gained by land- or space-based observations. One option is to establish an intensive, multi-year, surface-based observatory in the central Arctic Basin via an icebreaker-supported ice sta-tion. Such an ambitions endeavor will require the coordinated support of multiple U.S. agencies and international partners. The 1997-1998 Surface Heat Budget of the Arctic Ocean campaign (SHEBA), with participation by DOE, NASA, NOAA, and NSF was a good example of interagency coordination on a ship-based campaign that provided a first look into many central Arctic processes. Experience from SHEBA and other campaigns over the Arctic sea ice has shown that the most valuable data have been obtained from campaigns making comprehensive, multi-disciplinary measurements during one or more annual cycles; now in the "new Arctic" is an opportune time to look a second look at a substantially changed central Arctic.

Finally, operational models and model re-analyses assimilate observational data to improve model accuracy. The central Arctic Basin is, however, sparsely populated with routine observational inputs for these models. First, efforts should be made to improve operational satellite products through ground- and aircraft-based validation and algorithm development to provide improved constraints on model performance (NASA and NOAA). Second, expanded operational observations of basic Arctic atmospheric properties are needed, such as those provided by additional radiosonde stations or via routine dropsonde observations. UAVs offer great promise for cost-effective means to make sustained measurements over the central Arctic (DOE, NASA, and NOAA). These observational enhancements have the potential to reduce uncertainty in operational models and re-analysis products over both the Arctic and lower latitudes.

References

Darby, L. S., *et al.* (2011). International Arctic System for Observing the Atmosphere (IASOA). In I. Krupnik, *et al.* (Eds.), *Understanding earth's polar challengers: International Polar Year 2007-2008*, Edmonton, Alberta, Canada: University of the Arctic.

Jacob, D. J., *et al.* (2010). The Arctic Research of the Composition of the Troposphere from Aircraft and Satellites (ARCTAS) mission: design, execution, and first results, *Atmos. Chem. Phys.*, 10, 5191-5292.

McFarquhar, G. M., *et al.* (2011). Indirect and Semi-Indirect Aerosol Campaign: The Impact of Arctic Aerosols on Clouds, *Bull. Amer. Meteor. Soc.*, 92, 183-201.

Solomon, S., Qin, D., Manning, M., Chen, Z., Marquis, M., Averyt, K. B., Tignor, M., and Miller, H. L. (Eds.). (2007). Contribution of Working Group I to the Fourth Assessment Report of the Intergovernmental Panel on Climate Change. Cambridge, United Kingdom and New York, NY, USA Cambridge University Press.

Quinn, P. K., Bates, T. S., Baum, E., Bond, T., Burkhart, J. F., Fiore, A. M., Flanner, M., Carrett, T. J., Koch, D., McConnell, J., Shindell, D., & Stohl, A. (2008). The Impact of Short-Lived Pollutants on Arctic Climate. ed. AMAP. Oslo, Norway: Arctic Monitoring and Assessment Programme.

Serreze, M. C., & Barry, R. G. (2011). Processes and impacts of Arctic amplification: A research synthesis. *Global and Planetary Change*, 77, 85-96.

Uttal, T., *et al.* (2002). Surface Heat Budget of the Arctic Ocean, *Bull. Amer. Meteor. Soc.*, 83, 255-275.

3.4. Integrate and continue to deploy a national Arctic observing system and promote international cooperation to create a circumpolar Arctic observing system

Lead Authors:
C. Nikoosh Carlo, NSF
Simon Stephenson, NSF
Martin O. Jeffries, ONR
Robert Sanford, NSF

Agency Partners:
DHS, DOE, DOI, EPA, NASA, NOAA, NSF, ONR

The Arctic Observing Network (AON) is being developed to provide data streams for parameters key to understanding the changing Arctic environment. Initially, a U.S. activity conceived by the Study of Environmental Arctic Change (SEARCH) group coordinated Arctic observing efforts expanded internationally during the International Polar Year 2007–2008, and these international efforts are now endorsed by the Arctic Council. Several documents and reports provide a vision of what an AON should be, as well as recommendations for network design, data approaches, input from local and other stakeholders, response to agency specific needs, and international partnerships (see end note). International collaboration has been developed further under the umbrella of the Sustaining Arctic Observing Network (SAON), including participation by the eight Arctic countries, indigenous organizations recognized as permanent participants in the Arctic Council, Arctic Council working groups, and non-Arctic countries and entities (Calder 2011). The U.S. contribution to the World Meteorological Organization (WMO) Global Atmosphere Watch provides another international cooperation mechanism.

For a list of Federal observing efforts see, *Arctic Observing Network: Towards a U.S. contribution to Pan-Arctic Observing* (IARPC 2007).

This section focuses on areas where interagency collaboration will greatly advance Arctic observing priorities and support the development of coordinated and integrated regional, national, and global observing systems. It will not attempt to review ongoing efforts at individual agencies. Suggestions for the general direction of network development are:

1. Agencies should establish key activities or priorities for Arctic research to support their mission. For example, NASA's comprehensive approach uses satellite, aircraft, *in-situ*, and model output to understand the causes of large, rapid changes in Arctic sea ice, glaciers, ice sheets, land surfaces, permafrost, and atmospheric composition and chemistry. Those physical parameters strongly influence health and welfare of citizens throughout the pan-Arctic region, short-term weather patterns over the United States, and global change. NOAA's *Arctic Vision and Strategy* targets accurate, quantitative, daily to decadal predictions to support safe operations and ecosystem stewardship. Both priority areas are also included in the National Ocean Policy's Implementation Plan (http://www.whitehouse.gov/administration/eop/oceans/objectives) and will benefit greatly from interagency collaboration.

2. Residents of the North—individuals, communities, and their representative governments—need sound information and decision-making tools to inform adaptations to future conditions that will preserve, to the extent possible, their livelihoods, access to resources, and cultures.

3. Research should continue to build our understanding of the changing Arctic environment system and contribute to the design of an optimal observing network. From this, new priorities will be added; older ones may decline in importance or change focus.

The IARPC plans to focus on nine interagency observing network efforts over the next five years. A formal review and update of those priorities will occur every two years.

3.4.1 Facilitate observing-system design for the arctic

Why do this?

Long-term observations *in-situ*, from space, and by local people and communities have been vital to documenting changes occurring throughout the Arctic environment. Without remote-sensing observations from space, for example, the recent dramatic changes in Arctic Ocean sea-ice extent with melting and mass loss in the Greenland ice sheet with and changing tundra "greenness" would not have been detected and quantified quickly. The recognized need for a diverse set of pan-Arctic observations that would improve the value of predictive models spawned the AON. The SEARCH articulated the need for AON in the first part of the last decade and continues today to refine thinking about the system via the SEARCH AON Design & Implementation Task Force (see the 2010 AON Program Status Report). More recently, the need was discussed in the National Academy of Sciences report, *Toward an Integrated Arctic*

Observing Network. Components of the AON have been well-designed; the challenge now is how to link these effects across locations, regions, disciplines, or sectors.

Progress in understanding the changing Arctic environmental system will be achieved by a variety of methods, including the use of coupled, numerical models that represent a regional Arctic system. Ideally, observations vital to model initialization, calibration, assimilation, and skill-testing are best provided by an optimal network. Optimizing an observing network is a design exercise that itself would employ numerical models to establish observing priorities and identify gaps, optimal observing sites, and observational needs, such as variables to be observed and the frequency and duration of observations. Thus, observing-system design and Arctic-system modeling are symbiotic and require the observing and modeling communities to work together to identify and develop synergies that will improve both approaches. Arctic regional models are discussed in more detail in section 3.5.

3.4.2 Assess local-resident priorities for addressing change

Why do this?

Polar amplification of global environmental change is forcing local Arctic residents to address change immediately and to identify new adaptation strategies while balancing ongoing social, cultural, health, and economic demands and expectations (see sections 3.6 and 3.7). Priorities for the Alaska Native communities include: human health well-being, subsistence rights and access, development of alternative energy, and rural dependence on fossil fuels, including predicting how climate change will alter their environment (see Alaska Federation of Natives 2010 and 2011 Federal Priorities and the Inuit Circumpolar Council Alaska Strategic Plan 2010-2014). Some observing networks should enable informed decision making by local Arctic residents.

3.4.3 Combine *in-situ* and remotely sensed observations of sea ice with local community and traditional knowledge

Why do this?

The changing seasonality of sea ice will have profound climate, environmental, socio-economic, and political consequences. A sea-ice observing system that combines *in-situ* and remotely sensed observations with local community and traditional knowledge can meet the growing need for information about the state of the sea-ice cover at daily, weekly, monthly, seasonal, annual, and decadal temporal scales. As the Arctic Ocean moves toward a more maritime environment, the challenge will be to develop an optimal and integrated sea-ice observing system and derived information products and tools that will serve individuals and communities, industry, governments, and scientific research. The observing system will improve understanding of sea-ice processes and changing seasonality, operational forecasts, and longer-term predictions. Sea-ice processes, interactions, feedbacks, and modeling are discussed in more detail in section 3.1.

3.4.4 Conduct long-term monitoring of key outlet glaciers and tidewater glaciers

Why do this?

Long-term monitoring is the only way to observe and study the evolving relationships between climate changes, perturbations at the ice/ocean interface, glacier flow, and mass loss. Periodic measurements at key sites near Greenland and Arctic tidewater glaciers should capture glacier flow, local meteorology, ice mélange conditions, and oceanic conditions near the glacier front, in the fjord, and on the continental shelf. Data collected should also provide a measure of the heat and freshwater transport into and out of key fjords to enable heat budget analyses and provide boundary conditions for ocean general circulation models (GCMs). Essential variables include ice elevation, mass balance and flow speed, local meteorology, ocean temperature and salinity, and sea-ice conditions. These data will provide inputs for models of land ice loss described in section 3.5 as well as invaluable context for the study and validation of the linkages between key processes operating at vastly differing scales.

3.4.5 Monitor the biological and physical state of the Arctic marine environment

Why do this?

Unprecedented ocean warming and seasonal thinning and retreat of sea ice are altering the biology of the Arctic Ocean and could potentially lead to ecosystem reorganization. The international Arctic Council has initiated a program to monitor the biological state of the Arctic marine environment. In the United States, a DBO in the Pacific Arctic sector with emphasis on the Beaufort and Chukchi Seas is envisioned to provide biological and supportive environmental data at explicit regional sites. Portions of the DBO represent the U.S. contribution to the Arctic Council's monitoring program. The DBO would use a collaborative international network of logistical support to track ongoing shifts in ecosystem structure concomitant with climate change. These are discussed more fully in section 3.1.

3.4.6 Assess the effects of clouds and atmospheric constituents on surface radiation balance

Why do this?

While the effects of Arctic warming are easily observed, the energy sources driving those changes are poorly understood. The surface-energy balance—the amount of energy from the sun versus the amount reflected or remitted back into space—is largely unknown due to the unusual effects of clouds and aerosols in polar regions (see section 3.3). The half-year-long polar night, the highly reflective surface of ice, and the unique atmospheric stability allow clouds and aerosols to enhance either warming or cooling depending on conditions. Short-lived climate forcers, which include black-carbon aerosols, ozone, and methane, are also thought to enhance Arctic radiative forcing. These and other complex cloud-aerosol processes contribute to some of the largest uncertainties in Arctic regional climate models.

3.4.7 Assess the impact of terrestrial warming and permafrost thawing on the carbon cycle

Why do this?

Recent estimates show that soils of high-latitude ecosystems store about 2,000 petagrams of carbon; a catastrophic release of carbon from the soils could be many times greater than anthropogenic carbon dioxide emissions. Climate change scenarios predict the greatest magnitude of global warming will occur at high latitudes, and considerable observational evidence indicates recent warmer ground temperatures and permafrost thawing. The size of soil carbon pools and their sensitivity to temperature changes suggests a net loss of old-soil carbon to the atmosphere, causing a positive feedback to climate change. Terrestrial ecosystems are discussed in more detail in section 3.2, and marine ecosystems, including greenhouse-gas fluxes, are discussed in more detail in section 3.1.

3.4.8 Improve data access

Why do this?

Many data sets can be archived in the NASA's Earth Observing System Data and Information System (EOSDIS); NOAA National Data Centers (Geophysical, Oceanographic, and Climate Data Centers); NOAA's Arctic Ocean Observing System (AOOS); DOI/USGS Earth Resources Observation Systems (EROS) Data Center; DOE Atmospheric Radiation Measurement (ARM) and Next-Generation Ecosystem Experiments (NGEE) Data Archives; the NSF-supported Advanced Cooperative Arctic Data and Information Service (A-CADIS); the Geographic Information Network for Alaska at the University of Alaska; and the Exchange for Local Observations and Knowledge (ELOKA). Note that this is not an exhaustive list. The aim of these resources is to enable scientific discovery by allowing scientists to easily access, share, integrate, and work with data spanning multiple disciplines.

There remain several challenges that could benefit from interagency resources and expertise, which are detailed further in the Research Infrastructure Chapter 4, Table 3. These include, but are not limited to data exfiltration (unauthorized data transfer), data processing and archiving, establishing data standards, and data agreements that allow for interoperability of archives and support data discovery, analysis, and integration, and derived product development.

3.4.9 Engage indigenous observers and communities in monitoring environmental parameters

Why do this?

Federal agencies supporting various components of the emerging circumpolar Arctic observation system, both nationally and internationally, will continue to work toward integrating community-based observation networks with physical and biogeochemical monitoring systems. Introducing human-focused components to observing and monitoring Arctic change has helped move the AON/SAON network far ahead of previous efforts. It is now widely recognized that engaging indigenous observers and communities in direct monitoring of various environmental parameters, such as weather, sea ice, coastal erosion, water resources, permafrost, marine and riverine resources, and terrestrial wildlife provides considerable value to both communities and researchers. There are strong positive examples such as EALÁT, a Reindeer Herders Vulnerability Network Study; the Bering Sea Sub-Network (BSSN); and the Sea Ice Knowledge and Use (SIKU) study—and similar projects begun under the IPY 2007-2008 program. During the next five years, agencies will explore the viability of converting such local initiatives into longer-term activities.

Timeframe:

Long-term (>5 years) to achieve a sustainable AON; short-term (1-3 years) to make substantial progress on several high-priority areas.

Expected Outcomes:

- Improved understanding of how the Arctic system is changing and its connections to global systems.

- Development of coordinated and integrated regional, national, and global observing systems.

- Informed decision making by local Arctic residents and other stakeholders.

- Easy access to, sharing of, integration across data spanning multiple disciplines.

- Coordinated monitoring network, including space and airborne missions, to measure and monitor the overall controls on Arctic land ice mass change, and the freshwater input into the North Atlantic.

Milestones:

- Support diverse, multi-disciplinary observing teams that include representatives from state, local, and tribal governments, academia, the private sector, the international Arctic community, and other stakeholders (DHS, DOE, DOI, EPA, NOAA, NSF, ONR; FY2014).

  – U.S. Arctic Observing Coordination Meeting (DOI, NOAA, NSF, ONR; FY2012).

  – Arctic Observing Summit (NOAA, NSF, ONR; FY2013).

- Assess the state of the nine observing system/network themes and identify knowledge gaps and establish sites or regions for key observations (DHS, DOE, DOI, EPA, NOAA, NSF, ONR; FY2013).

- Develop an interagency (and international) planning document for an Arctic land ice monitoring system with focus on outlet and tidewater glaciers and their surrounding. (DOE, NASA, NOAA, NSF; FY2015).

- Develop action plans to implement an integrated design, including connections with other national and international observing systems, sustain current and planned operations, and use system models to identify observing contributions and needs for forecasting and design (all agencies; FY2016).

Science and Technology Gaps and Needs:

New activities under this plan should be informed by ongoing efforts at the local, tribal, state, Federal, and international levels. Efforts should be made to review existing activities and create synergisms when goals and priorities align. Current research efforts would benefit from advisors from across Federal, state, local, and academic research organizations and industry to address AON topics. Such diverse perspectives on high-priority topics could quickly mobilize resources to advance areas of critical need. Current inability to find some key datasets argues for a common portal to assist data discovery. Such a portal should be accessible and intuitive for local community members, educators, and the scientific community alike. Better access is needed to radar imagery over sea ice, glaciers, ice caps, ice sheets, and permafrost regions. The ESA Sentinel series and the Canadian Space Agency Radar Satellite (CSA RADARSAT) Constellation will partly fill this gap but will depend on partnerships with overseas operations including free and open data sharing. Monitoring land ice loss will require continued development and deployment of an interdisciplinary observing system that is low-energy consuming, weatherproof, highly telemetrically efficient, and robust. In general there is great need for real-time data analysis systems capable of dealing with different data sources, types, and formats; and space and airborne campaigns that provide a continuous observational record.

References

Alaska Federation of Natives. (2010) Federal Priorities. http://www.nativefederation.org/AFNFedPriorities.php.

Arctic Observing Network [AON]. (2010). *Arctic Observing Network (AON) Program Status Report*. Results from the Third AON Principal Investigators (PI) Meeting, 30 November–2 December, 2009, Boulder, CO. http://www.arcus.org/search/aon.

Calder, J. (2011). Sustaining Arctic Observing Networks (SAON). In Krupnik, I., *et al.*, (Eds.), *Understanding Earth's Polar Challenges: International Polar Year 2007-2008*. Edmonton, Canadian: Circumpolar Institute.

Interagency Arctic Research Policy Committee [IARPC]. (2007). *Arctic Observing Network: Towards a U.S. Contribution to Pan-Arctic Observing*. http://www.nsf.gov/pubs/2008/nsf0842/index.jsp.

Inuit Circumpolar Council. (2011). *Inuit Circumpolar Council Alaska Strategic Plan 2010-2014*. http://www.iccalaska.org/servlet/content/other.html.

National Research Council. (2006). *Toward an Integrated Arctic Observing Network*. Washington, D.C.: The National Academies Press. http://www.nap.edu/catalog.php?record_id=11607.

Schlosser, P., Tucker, Warnick, W., & York, A. (Eds.). (2003). *Arctic Research Support and Logistics: Strategies and Recommendations for System-scale Studies in a Changing Environment*. Fairbanks, Alaska: Arctic Research Consortium of the United States. pp. 81. http://www.arcus.org/logistics/03_report.html.

Solomon, S., Qin, D., Manning, M., Chen, Z., Marquis, M., Averyt, K. B., Tignor, M., and Miller, H. L. (Eds.). (2007). Contribution of Working Group I to the Fourth Assessment Report of the Intergovernmental

Panel on Climate Change. Cambridge, United Kingdom and New York, NY, USA Cambridge University Press.

Study of Environmental Change [SEARCH]. (2005). *Study of Environmental Arctic Change: Plans for Implementation During the International Polar Year and Beyond*. Fairbanks, Alaska: Arctic Research Consortium of the United States (ARCUS). http://www.arcus.org/search/resources/ reportsandscienceplans.php.

End note:

Additional background on AON appears in:

- 2003 Arctic Research Support and Logistics: Strategies and Recommendations for System-scale Studies in a Changing Environment. ARCUS

- 2005 Study of Environmental Arctic Change (SEARCH): Plans for Implementation During the International Polar Year and Beyond. ARCUS

- 2006 Towards an Integrated Arctic Observing Network. NAS

- 2007 Arctic Observing Network: Towards a U.S. contribution to Pan-Arctic Observing. IARPC

- 2009 Arctic Observing Network (AON) Program Status Report. Results from the Third AON Principal Investigators (PI) Meeting, 30 November–2 December, 2009, Boulder, CO

3.5. Integrate Arctic regional models

Lead Author:
Mike Kuperberg, DOE

Agency Partners:
DOE, DOI, NOAA, NSF, ONR

Models of Earth's climate are mathematical tools for understanding climate processes and their feed-backs, as well as for predicting and projecting climate variability and change. A variety of models are being applied in the Arctic region for projecting future climate change, forecasting Arctic weather and sea-ice conditions, and understanding Arctic processes. Yet, due to large feedbacks, sparse observations, critical gaps in process understanding, differences in process representation, and high variability of vegetation, ice, and snow cover, significant uncertainty exists in simulations of Arctic changes. An integrated and focused effort to improve Arctic models would benefit understanding of ongoing processes, ability to project future Arctic changes, and informed use of those projections. Strongly coupling modeling and process-science research ensures that models incorporate state-of-the-science knowledge about critical systems. Process studies (e.g., observations, experiments) can advance understanding in weak or unrepresented areas, and the resulting improved models can be used to guide field research/labo-ratory research and to inform future decisions. Parameterization is a challenge of scaling—describing complex, smaller-scale processes as a simplified, mathematical representation in larger scale models. Close and improved collaboration between modeling and process scientists is needed to develop, test, and implement parameterizations. The climate community has produced a number of documents that propose research directions; the following sections have made significant use of the *Science Plan for*

Regional Arctic System Modeling (IARC), *Modeling and Predicting Arctic Weather and Climate* (Second *International Conference on Arctic Research Planning* (ICARP II), Science Plan 9, and the Arctic Climate Impact Assessment (ACIA).

3.5.1 Inventory Arctic modeling activities

Why do this?

Significant efforts have been made to coordinate and share research on Arctic regional modeling. Rapid changes in the pan-Arctic region, however, call for even greater coordination of agency efforts and planned scientific research in Arctic change and modeling. Existing sources and portals that provide a base for comprehensive information and data coordination and sharing include:

- National Snow and Ice Data Center (NSIDC), Boulder, CO, which includes NASA's Distributed Active Archive Center (DAAC) for all cryosphere related data and NSF's Advanced Cooperative Arctic Data and Information Service (ACADIS).

- The Arctic Portal established as part of the International Polar Year activities and currently maintained by Iceland and various Arctic Council Working Groups

- ARM Climate Research Facility Archive (http://www.archive.arm.gov/armlogin/login.jsp)

- Earth System Grid (http://www.earthsystemgrid.org)

- Polar Hydrography Center, University of Washington

- U.S. National Academies Polar Research Board

- Alaska Ocean Observing System (http://www.aoos.org/)

- North Slope Science Initiative (http://northslope.org/)

- Arctic Ocean Model Intercomparison Project workshops

- NASA Catalog of Earth Science System Components (http://www.asd.ssc.nasa.gov/m2m)

- Community Earth System Model (CESM) Polar Climate Working Group (http://www.cesm.ucar.edu/working_groups/Polar)

- NSF's Arctic Long Term Ecological Research (LTER) sites (http://www.lternet.edu/)

- Arctic Landscape Conservation Cooperative (http://www.arcticlcc.org)

- Scenarios Network for Alaska Planning (http://www.snap.uaf.edu/)

Access to and broader use of existing models, data systems, and information portals would better coordinate and integrate existing agency efforts and improve dissemination of data and information about Arctic research activities. An inventory of modeling activities, process research, and data sources would provide a basis for defining collaborative programs and coordinated campaigns for advancing Arctic modeling. Cataloging the existence and nature of activities, such as model inter-comparisons and benchmarking, should be included in such an inventory.

Timeframe:

Near-term (1-2 years)

Expected Outcomes:

- New efforts to intrinsically couple modeling and process research activities.

- More rapid integration of knowledge into models.

- Improved capability for model evaluation to inform process research.

Milestones:

- Conduct and disseminate a survey of Federal Arctic modeling efforts (DOE, NSF; FY2013).

- Evaluate the results of the survey and identify opportunities for collaborative development and/ or joint campaigns (DOE, DOI, NASA, NOAA, NSF, ONR; FY2014).

Science and Technology Gaps and Needs:

A comprehensive bibliographic database of results from international entities [such as the International Arctic Science Committee and the World Climate Research Program's (WCRP) Climate and Cryosphere Program] would provide needed context for how modeling relates to ongoing studies.

3.5.2 Encourage coordinated approaches that better represent Arctic processes in Earth-system models

Why do this?

Global Earth system models (ESMs) have become increasingly robust, accurate, and comprehensive. Large differences exist in the Arctic systems simulated by different ESMs and many processes are either absent (dynamic ice sheets, ice ecosystems), poorly represented (multi-phase clouds, permafrost, vegetation) or unresolved (ice fracturing, ice shelves, ocean eddies) in Arctic models. Feedbacks in the Arctic are also particularly strong and sensitive; thinning sea ice is very sensitive to both atmosphere and ocean forcing and becomes highly variable with subsequent changes in albedo and other feedbacks; biogeochemical processes that control the rate and nature of carbon release are suspected to lead to dramatic effects on atmospheric carbon concentrations from thawing permafrost; and ecosystem changes may have direct (albedo) and indirect (ecosystem services) impacts.

Such processes and feedbacks control both regional Arctic and, indirectly, global climate variability. Their realistic representation requires models with very high spatio-temporal resolution as well as detailed and often long-term observations to improve parameterizations and model verification. High-resolution models of Arctic processes will subsequently need to be parameterized and/or scaled for use in current ESMs. International coordination of efforts and cost sharing are important while maintaining model diversity for more robust evaluation of a variety of approaches.

Timeframe:

Mid-term (3-5 years) for many model improvements already in the pipeline; long-term (5-10 years) for next-generation models. The Earth System Modeling Framework (ESMF) (http://www.earthsystem modeling.org/) and the Earth System Grid Federation (ESGF) (http://esgf.org/) are currently making models and their output accessible to the larger community but further coordination is needed.

Expected Outcomes:

- Improved interagency coordination through dedicated multi-agency campaigns on specific Arctic processes.

- Continued development of common modeling frameworks, data and meta-data standards to facilitate increased sharing of model components, simulation, and observational data.

- Continued improvement of process representation of the Arctic region in Global Earth system models.

Milestones:

- Identify critical Arctic processes for dedicated field and modeling campaigns across agencies (DOE, DOI, NASA, NOAA, NSF, ONR; FY2015).

- Coordinate Federal activities to develop, implement, and test improved parameterizations of Arctic physical processes and feedbacks (DOE, DOI, NOAA, NSF, ONR; FY2015).

- Develop standardized model components, meta-data, and data products (DOE, DOI, NOAA, NSF; FY2016).

- Conduct model inter-comparisons to foster collaboration among modeling groups and identify high-priority Arctic model improvements (DOE, NOAA, NSF; FY2018).

- Review report from the third session of the WMO Executive Council Panel of Experts on Polar Observations, Research, and Services on the development of the Global Integrated Polar Prediction System (NOAA, NSF; FY2012).

Science and Technology Gaps and Needs:

While some coordinated efforts exist, such as the Arctic Ocean Model Intercomparison Project (AOMIP) and Climate Process Teams, additional opportunities exist to leverage and coordinate current and planned Arctic research investments. Enhanced mechanisms for interagency collaboration are therefore needed.

3.5.3 Build Arctic and subsystem models for coupling with regional and global approaches

Why do this?

Regional models of the Arctic are important tools that enable explicit representation of fine-scale atmospheric, oceanic, terrestrial, and cryospheric processes at scales where they occur. Regional Arctic models can also take advantage of the growing number of data streams and process understanding that is emerging from mechanistic studies and field and laboratory observations. They can also highlight the most critical and sensitive processes and parameters for inclusion in regional and global models.

Regional models are required to fully incorporate the complex interactions between sea ice, ice sheets, cold oceans, regional climate, permafrost stability, and their integrated influence on the regional and global carbon cycle and energy balance. Ultimately, the understanding obtained from these models can be incorporated into global models either through coupling/nesting or by creating appropriate parameterizations. Global Earth system models will be the primary tool for exploring climate change and the Arctic feedbacks in the global climate system.

Timeframe:

Mid- to long-term (3-10 years). Focused development of new subsystem models and interaction with experimentalists (>5 years); coupling and evaluation within regional and later in global models (5-10 years).

Expected Outcomes:

- A hierarchy of models from process to regional scales.

- Inclusion of Arctic high-resolution physical, biological, chemical, and social subsystem model components within a regional Arctic climate system model.

- Improved representation of the Arctic system in Earth system models through nesting of regional models or development of parameterizations.

Milestones:

- Develop and evaluate stand-alone subsystem components of the Arctic system, incorporating mechanistic processes derived from experiments and/or observations (DOE, DOI, NSF; FY2014).

- Couple, test, and validate the aforementioned models against observations of subsystem components within a regional Arctic climate-system model (DOE, DOI, NOAA, NSF, ONR; FY2016).

- Couple and evaluate Arctic subsystem components within Global Earth system models (DOE, DOI, NOAA, NSF, ONR; FY2017).

Science and Technology Gaps and Needs:

Observational data are needed to constrain and rigorously evaluate individual subsystem model components and multi-component interactions and dependencies. Advanced *in-situ* and remote platforms are needed for coordinated process studies, especially for difficult observations such as under-ice sampling and communication, melting out/freezing in, long-time/distance endurance, high-resolution, large-scale, and long-term coverage. Also, community participation in model development and data collection could be better supported by organizational structure as would common, coordinated Arctic data/model distribution center(s) for community posting and use of models.

3.5.4 Develop models of Arctic land ice mass loss, connections to ocean and atmospheric variability, and implications for sea level

Why do this?

Predicting future changes in ice sheet mass and the impacts on sea level, ocean circulation, and climate will require improved understanding of ice-sheet and ice-ocean processes. Data from process-oriented studies, targeted campaigns, and long-term monitoring systems must be integrated into large-scale circulation and Earth-system models. The spread of current projections of sea level rise contributions from Greenland by 2100, from 0.06 to 0.54 m, with an added 0.34 m from glaciers and ice caps, gives a crude measure of current uncertainties.

Timeframe:

Mid-term (3-10 years)

Expected Outcomes:

- Implement and test physically based parameterizations of unresolved processes: key physical processes operating on scales of 100 m or less need to be incorporated into Earth system models using a suite of parameterization techniques.

- Comprehensive, well-structured, and sophisticated databases and data formats which allow rapid access and optimal use of the hard-won data.

- Coupling of the various components of the Earth system models: representation of feedbacks between ice sheet variability and the large-scale ocean/atmosphere circulation or other climate system components through interactive (two-way) coupling between ice sheet and climate models or components.

- Coupling and accounting for Earth gravity/geodetic responses in models: glacial isostatic adjustment (GIA) from visco-elastic mantle response to unloading and gravitationally self-consistent sea-level solutions are crucial in assessing regional sea level change as a consequence of ice sheet mass loss; gravity models of improved accuracy are needed and results fed back into Earth system models (which currently neglect these effects).

- Model testing, analysis, and intercomparison: a hierarchy of modeling approaches serves as a quantitative basis for model assessment and identification of systematic biases; a feedback loop links large-scale model-data biases, e.g., subpolar gyre hydrographic properties, to those at key locations (fjord exits and glacier termini), and to discrepancies of parameterized versus observed processes.

- Use of models to optimize observing-system development: these studies will assess which processes have the strongest impact on constraining ice mass loss, and where, with what accuracy, and at which frequency these should be sampled.

Milestones:

- Support surveys of existing modeling capabilities by participating agencies that tackle the ocean/land ice coupling problem, including parameterization and moving geometry approaches (DOE, NOAA, NSF; FY2014).

- Help coordinate the implementation of multi-disciplinary test suites and scenarios for idealized and realistic model verification, validation, and intercomparison (a combination of ISOMIP and SeaRISE but extended to the coupled problem); foster exchange of codes, inputs, experiences (DOE, NSF; FY2014).

- Translate results from process studies into parameterizations for use in Earth System Models (DOE, NOAA, NSF; FY2015).

- Development of new-generation modeling approaches into the coupled forward and inverse problem (DOE, NOAA, NSF; FY2018).

- Completion of observing-system studies using improved models to further inform observing-system design (DOI, NOAA, NSF; FY2023).

Science and Technology Gaps and Needs:

- Numerical methods are needed that can deal with processes acting on a wide range of spatio-temporal scales, from sub-centimeter boundary layers to ocean basin-scale (1000 km) coupled atmosphere/ocean dynamics. Three types of complementary approaches may be pursued: (1) process-resolving simulations to develop process parameterizations; (2) feedback and testing of these parameterizations in large-scale Earth system models; and (3) in parallel with (1) and (2), the development of new computational tools (unstructured and/or adaptive meshing, multi-grid, multi-physics approaches) to enable seamless simulations. Algorithms are required that can efficiently exploit emerging massively parallel computer systems and integrate different system components. Yet, simulations also need to produce diagnostic output that is useful for scientific analysis.

- Dynamical feedbacks through moving geometry (glacier retreat) represent a new computational challenge in climate modeling; coupled ocean-glacier simulations that aim to capture dominant glacier stress perturbations through glacier un-grounding and retreat need to represent the changing interface geometry.

- Observational sparsity requires the continued development and use of formal inverse/estimation methods in a coupled model/data framework; merging of diverse observations and models is needed to infer optimal state and parameters and to assess remaining uncertainties.

3.5.5 Increase Arctic model resolution to improve prediction and inform future research and observations

Why do this?

Accurate simulations of the Arctic require the ability to resolve features and processes at small spatial and temporal scales. For example, Arctic Ocean eddies, cloud processes, ice fracturing, ice-shelf dynamics, ecosystem/vegetation changes, and hydrology all occur on kilometer scales or smaller. Through data synthesis and integration with other models, high-resolution regional models can guide future observations and process studies while providing a large-scale and long-term picture of Arctic system variability. Variable resolution approaches are now being implemented and show great promise for enabling high-resolution Arctic models in a global model framework. Similarly, alternative time integration techniques are being explored to capture multiple time scales consistently and accurately.

Process understanding, data sets, and model representations are needed at small spatial and temporal scales. High-resolution model output is also commonly required by stakeholders, especially for impact assessment and mitigation/adaptation studies. While high-resolution global models are beginning to reach these spatial scales, continued use of regional, stand-alone, and nested models will be needed to provide ensemble simulations at the required resolutions. Similarly, fast processes related to surface exchange and boundary layer processes must be represented, as well as decadal modes of variability and longer processes like ice-sheet melting, ecosystem evolution, and ocean thermohaline circulation—all of which have significant impacts on the Arctic. Results from those simulations can inform process research direction and observing-system design (see section 3.4). Already, high-resolution simulations have been used to help design large-scale ecosystem studies in Arctic permafrost and float experiments in the Southern Ocean.

Timeframe:

Mid-term (3-5 years) for improved Arctic climate-system models and variable-resolution models for focused Arctic simulations

Expected Outcomes:

- Improved and detailed forecasts of Arctic climate change to address stakeholder needs.

- Arctic simulations at fine-enough resolution to inform experimental/observational design and field campaigns.

Milestones:

- Conduct ensemble simulations of future Arctic climate change at kilometer spatial scales (DOE, DOI, NSF; FY2018).

- Engage the Arctic (and Earth system) modeling community in planning and designing future field campaigns (DOE, DOI, NOAA, NSF, ONR; FY2012).

- Provide mechanisms for rapid access to processed (quality controlled, formatted/gridded) observational data sets for model-data inter-comparison (DOE, NOAA, NSF; FY2015).

Science and Technology Gaps and Needs:

Much observational data are not at high enough resolution to validate kilometer-scale models or to incorporate into model-data inter-comparisons. Important gaps in process understanding limit full representation of the Arctic in coupled models. Advanced numerical techniques are needed for multi-scale time integrations. High-performance computing and technical support are needed for model integration thereby enabling co-design of model algorithms and/or new modeling approaches to carry out high-resolution simulations on new computing architectures.

3.5.6 Use insights from models to inform process research; use process research to evaluate and improve models

Why do this?

Models can be used to simulate processes that are difficult to observe and, thus, poorly understood. Analyzing models can highlight areas where new observations or studies will improve representation of processes in models and ultimately benefit predictions about Arctic climate change. Modelers are eager for robust data sets to evaluate and test their products, and experimentalists are collecting observations and data that can be incorporated into models. The two communities, however, do not often interact directly.

The disconnect among modelers, experimentalists, and observers has led to gaps in both the completeness of data for good prediction and in critical process representation in models. Coordinated research efforts are particularly needed in the Arctic where physical conditions and remote locations make it difficult to conduct process research and where different models currently provide very different representations of Arctic parameters and/or processes. Great opportunity exists to rapidly improve the design of Arctic experiments and observations, which would enable enhanced parameterization of subsystem, regional, and global models. Such experiments and observations also could identify and address structural uncertainty in models by including coupled experimental and predictive science components to identify gaps and needs simultaneously for both data and models.

Timeframe:

Near- to long-term (1-10 years). Current field-research efforts and laboratory process studies are already being informed to some extent by model analyses, but more comprehensive and integrated model and field campaigns will be needed.

Expected Outcomes:

- Improved understanding and model representation of processes specific to the Arctic system.
- A new cycle of systems model development and improvement, incorporating integrated field campaigns to develop refined parameterizations for Arctic process representations in models.
- More efficient use of research funds through enhanced collaboration among process researchers, observationalists, theoreticians, and modelers.
- Rapid increase in development of robust data sets, process understanding, and improved models and predictive skill.

Milestones:

- Publish SeaRise ice sheet model intercomparison results (DOE, NASA, NSF; FY2013).

- Analyze model output to determine future needs for data collection and process studies (DOE, DOI; FY2014).

- Design and implement integrated modeling and field campaigns focused on specific high-priority processes to improve process understanding and representation in models (DOE, DOI, NASA, NOAA, NSF, ONR; FY2012).

- Develop, test, and evaluate new/improved parameterizations (DOE, NOAA, NSF, ONR; FY2015).

- Plan and conduct interagency and intra agency conferences, workshops, and campaigns designed to bring process researchers and predictive scientists together to solve Arctic grand challenges (DOE, DOI, NASA, NOAA, NSF, ONR; FY2015).

- Develop funding solicitations that require integrated process-prediction research approach (DOE, DOI, NSF, ONR; FY2013).

Science and Technology Gaps and Needs:

Closely connecting process and modeling science is needed to rapidly improve and validate our ability to project future Arctic conditions. Many processes in Arctic systems are not well understood and could be targets for focused research. Examples of key areas of research needs include:

1. Ice thickness distribution and ice production due to ice deformations;

2. Multi-phase ice, brine channels, melt ponds, and water/brine transport;

3. Ice-sheet and ocean interaction, ocean circulation in tidewater fjords, and property exchange over the sill;

4. Time-dependent evolution of surface, englacial, and subglacial hydrology for the Greenland ice sheet;

5. Ocean circulation and water-mass formation in ice-free Arctic conditions;

6. Ocean estuary/fjord subsystem model component for use in climate models;

7. Land/submarine permafrost distribution and interaction with atmosphere and/or ocean;

8. Permafrost hydrology and changes in morphology and sheet flow and riverine transport of surface water, sediments, nutrients, and contaminants;

9. Arctic biogeochemical cycles within permafrost, sea ice, atmosphere, and ocean, including carbon, methane, ice algae, aerosol-cloud interactions, aerosol deposition on ice;

10. Arctic mixed-phase clouds and other cloud microphysical processes, including ice nucleation;

11. Surface exchange processes at ice/ocean, ice/atmosphere, and ocean/atmosphere boundaries;

12. Rates and ranges of change for plants, animals, and ecosystem function; and

13. Permafrost-soil-vegetation interactions.

3.5.7 Integrate Arctic climate-model results with observational validation and verification to understand the principal drivers and uncertainties of Arctic climate changes

Why do this?

Unified, regional Arctic climate-system models are ideal tools to integrate data and information across multiple components of the Arctic system. Such models may also be applied to advanced, probabilistic decadal projections. Models must be verified against analytic test cases and reference solutions and validated against the historical record. Ensemble simulations are required to sample the solution space for uncertainty quantification as well as to separate natural or internal modes of variability from externally forced changes.

As with all aspects of climate modeling, critical evaluation needs to continue for Arctic regional models. Our ability to validate model projections (and to describe the uncertainties associated with those projections) is critical to our ability to use modeling results. Model validation contributes to the optimal synthesis and integration of limited process understanding and observational data to advance Arctic science, including interdependencies across the Arctic system components.

Timeframe:

Mid- to long-term (3-10 years)

Expected Outcomes:

- Projection of future Arctic climate change and its impacts including variability estimates.

- Determination of key principle drivers of Arctic climate variability and trends.

- Integration of comprehensive analyses of causes and effects of Arctic climate change.

Milestones:

- Coordinate model experiments and inter-comparisons to critically evaluate regional model results against observations (DOE, NOAA, NSF; FY2014).

- Develop and implement standards for gridded observational data sets (DOE, NOAA, NSF; FY2014).

- Implement a common data portal for both observational data and common model experiments and projections (DOE, DOI, NSF; FY2018).

Science and Technology Gaps and Needs:

Improved diagnostic measures suitable for both observations and model data are needed, as well as statistical techniques for uncertainty quantification. Improved coverage and long-term monitoring of the Arctic will generate better knowledge of present climate and initial conditions for future projections. New techniques for filling in sparse observational data from the historical record will improve model validation. Model improvements and additions should occur as a continuous, iterative cycle. Enhanced computational and data infrastructure would enable improved integration of currently disparate data streams and the ability to exercise increasingly complex (and, therefore, realistic) model representations of Arctic systems. The enhancements include improved data archives and distribution, advanced informatics, and access to high-performance computing for high-resolution ensemble simulations.

3.6. Assess strengths and vulnerabilities of Arctic communities facing the impacts of climate change and assist in developing adaptation strategies and tools to maximize sustainability, well-being, and cultural and linguistic heritage

Lead Authors:
William Fitzhugh, SI
C. Nikoosh Carlo, NSF
Igor Krupnik, SI
James Partain, NOAA

Agency Partners:
DOI, DOS, EPA, NSF, NOAA, SI

Rapid Arctic change is forcing residents to adapt to new conditions created by environmental change and diverse socio-economic stressors. Age-old traditional responses, such as diversification of natural-resource harvesting, relocation, and dispersal or concentration in the resource-rich niches, offer valuable lessons, but by themselves may be inadequate to fully address challenges such rapid changes and modern economic transitions present. New community-based participatory research to identify regional and local vulnerabilities and adaptation tools contribute to the growing body of knowledge needed for decision-support science and policy recommendations. Understanding the impacts of climate change could be better served by adding base—or "bottom-up"—assessments of the current drivers of social well-being to computer-generated models and other top-down scenarios used in ecosystem modeling. Focused studies of major social parameters of change are under way that are national and international in scope. Integrative multi-disciplinary programs, like SEARCH, that seek to balance societal needs with research priorities advanced by rapid environmental change can help explore effective ways of engaging communities and Tribes in the issues of utmost public urgency, such as resource development or management of living resources. *Arctic Social Indicators II* and *Arctic Human Development Report II*, projects endorsed by Arctic Council, will emphasize the key role of local communities in both basic-science discovery and in creating adaption tools that are practical and efficient at the local, regional, and broader circumpolar levels.

Knowledge about thresholds and prospective breaking points, as well as the inherent strengths in the social fiber that affect community resilience, would be invaluable in formulating Federal agency policies and in forging collaborations with diverse local stakeholders. The overall scope of new research should be broad and encompassing, but the informative case studies are to be conducted at the community level.

Melding the historic effects of cash and natural-resource-harvesting economies with added stressors associated with modernization and ongoing climate change is challenging but necessary. The emphasis should be to provide Arctic residents with basic scientific knowledge necessary for community leaders to develop sustainable pathways for successful adaptation amid rapid environmental change and a variety of other stressors—while juggling diverse Federal, state, and local interests. Providing such knowledge will help preserve cultural vitality and improving health and overall well-being for the long-term. The need to support tribal communities so they have representation in climate change-related activities and can develop and implement climate change strategies has also been recommended by the Climate Change Adaptation Task Force in its 2011 Progress Report.

3.6.1 In collaboration with local communities, develop methods for assessing community sustainability and resilience and determine the efficiency of current adaptation strategies

Why do this?

Native communities have thrived in the Arctic for millennia, but today's challenges are occurring at a pace faster than ever and may be beyond the capacity of traditional adaptation strategies. As primary stakeholders that will be affected by climate change, Arctic communities and Tribes need useful, access-ible, location-specific information on changes that are occurring and how those changes will impact human well-being. There is an urgent need for effective methods to assess community sustainability as northern residents face various socioeconomic challenges, changing local infrastructure needs, and high transportation and living costs. Traditional ecological knowledge should be actively incorporated in future planning, and more research is needed to identify key indicators of successful adaptation at local and regional scales.

Timeframe:

Near-term (3-5 years)

Expected Outcomes:

- Inform new understanding on how climate change, past and present, has been/is being met by Arctic communities and how communities have adapted.

- Improve research and assessment tools for studying the sources of community vulnerability to climate and environmental change and implications for community adaptability developed in partnership with local communities and organizations.

Milestones:

In collaboration with other Arctic nations:

- Identify and develop a database on past and current adaptation strategies used by Arctic communities to combat climate change impacts (DOS[10]; FY2013).

- Determine which strategies have been most successful (DOS; FY2013).

- Document unintended consequences of previous strategies and responses to change (DOS; FY2013).

- All will be explored by the Adaptation Actions for a Changing Arctic Report (DOS; FY2013).

---

10. This, and subsequent references to DOS in this section, indicates DOS's role as the Federal lead in international Arctic Council activities, in particular, DOS's role as head of delegation to the Sustainable Development Working Group (SDWG).

Science and Technology Gaps and Needs:

Limited work has been done to identify what Arctic communities have done, or are currently doing, to adapt to the effects of climate change. Even less is known about the efficacy of those adaptation strategies. Research is needed to, first, identify the suite of climate-change adaptation strategies currently being (or that have been) used and, second, to analyze the efficacy of those strategies. Particular attention should focus on both community specific strategies and those that were applied more broadly at the regional, national, and circumpolar scales. All research needs to be conducted as joint efforts and in close collaboration with local Arctic communities.

3.6.2 Identify the current vulnerabilities of Arctic communities and ecosystems to climate change and explore their interactions with socio-economic and other stressors

Why do this?

The Arctic region is warming quickly. The impacts of this change on Arctic communities are occurring in conjunction with other stressors, such as high cost of living, infrastructure maintenance, lack of employment opportunities, and shortage of resources for community development (see Section 3.7 for discussion of health-related challenges). These stressors need to be identified and quantified in pragmatic terms. Successful adaptation measures will require attention to multiple stressors simultaneously as well as close engagement with communities to empower self-sufficiency, initiate smart development, and build adaptive capacity.

Timeframe:

Near-term (1-3 years)

Expected Outcomes:

- Assist local communities in better understanding current vulnerabilities to climate change.

- Strengthen or create new partnerships among local stakeholders and resource managers to assess the vulnerability of individual Arctic communities and develop locally based adaptation and more general mitigation strategies.

- In collaboration with local stakeholders, develop new understanding of how changes in Arctic ecosystems and climate will affect natural-resource harvesting practices.

- Improve public understanding of how the effects of climate change interact with socio-economic stressors in the Arctic.

- Advance engagement of Federal agencies with local communities in sharing information on management and use of terrestrial and marine resources, and support empowerment for the co-creation of knowledge.

- Community profiles that highlight impending vulnerabilities and indicators of resilience.

Milestones:

- Establish research and community collaborations aligned with local priorities and needs, including in planning, data collection, conceptualization, and interpretation of research results and recommendations (NSF; FY2013).

- Assess vulnerability of Arctic communities and ecosystems to climate change and socio-economic stressors:

  - Environmental Studies Program Social Indicators in Coastal Alaska-Arctic Communities Study (BOEM; FY2014).

  - Arctic Sustainability (ArcSEES) initiative (BOEM, EPA, NSF, USFW, USGS, and a consortium of French science funding agencies; FY2013).

- In collaboration with other Arctic nations, develop a standardized set of quantifiable socio-economic indicators of vulnerability in the Arctic Council's Arctic Social Indicators II Study (DOS; FY2012).

Science and Technology Gaps and Needs:

Research on the vulnerabilities and indicators of resilience of individual Arctic communities to climate change is limited. Thus, there is a need to develop quantitative measures of vulnerability and resilience to climate change for coastal communities, which can be compared with concurrent socioeconomic stressors. More-objective analysis will improve the content and quality of socioeconomic-impact assessments associated with policy actions and marine resource-management decisions. Such research will enable affected communities to develop strategies for best responding to the challenges and opportunities they face.

3.6.3 Develop projections of future climate scenarios and demographic conditions to forecast potential strengths and weaknesses of human and ecological systems in the Arctic

Why do this?

Given current global environmental changes, the future will not be like the past; scenarios of future conditions help define alternate projections of environmental and socioeconomic states and—with identified uncertainties—will be valuable to adaptation planning.

Timeframe:

Long-term (>5 years)

Expected Outcomes:

- Improved models that allow communities and resource managers to better assess potential community and ecosystem-level vulnerabilities to medium- and long-term climate scenarios as well as the relative benefit of specific actions to address those vulnerabilities at the decadal scale.

Milestones:

- In collaboration with other Arctic nations, develop a standardized set of socio-economic indicators to measure future community resilience (2020 and 2030 projections), including input of local resources, population fluctuations and migration, communication networks, and capacity to adapt via the *Arctic Council's Arctic Social Indicators II Study 2012* and *Arctic Resilience Report* (DOS; FY2015).

- Link climate models with projections of ecological and socio-economic change that include community dependence on harvesting local food sources (BOEM, DOE, NOAA; FY2017).

- Test existing scenarios of the status of certain wildlife species of value to local consumers, according to available climate change models (USFW; FY2017).

Science and Technology Gaps and Needs:

Knowledge is lacking about how climate change might disproportionately impact communities and the ecosystems that they depend on for survival. A comprehensive research plan needs to be developed to collect the suite of information needed to model forecasted impacts. Once such models are created, adaptation strategies that respond to those impacts can be developed with local communities and various state and Federal agencies that can aid in moving explicit goals forward.

3.6.4 Assist Arctic communities in documenting, revitalizing, and strengthening indigenous languages and cultural heritage

Why do this?

Arctic communities have long faced threats to their indigenous languages and cultural continuity. New technologies, communication, education, and digital media are primarily in non-indigenous languages. At the same time, the digital era offers new channels for distant education, online publication, and broad dissemination of cultural materials. It is also creating new challenges, as most traditional cultural, subsistence, and language skills are still being transmitted by direct contact within families and communities. The impact of these and other new forces on the Arctic's indigenous people is poorly understood. Federal institutions need to develop a unified strategy or policy for assisting Arctic residents with 21st century challenges to their cultural well-being. Many communities are eager to address the progressive loss of traditional knowledge and language and are willing to work in partnership with Federal and local agencies to develop new strategies to preserve and use their unique cultural heritage.

Timeframe:

Mid-term (3-5 years)

Expected Outcomes:

- Assess the existing Federal and State of Alaska resources that support Arctic people's cultural heritage and ways to expand collaboration with indigenous communities in language, knowledge, and heritage preservation.

- Enable informed decision making by users—local programs, cultural institutions, schools, branches of local governments and communities—for collaborative actions in support of indigenous cultural heritage, languages, and identities.

- With local communities develop new programs and templates for cultural preservation, including language revitalization.

Milestones:

- In concert with local communities, strengthen partnerships between researchers, Alaska Native organizations, Federal, State, and non-governmental organization (NGO) entities through strategic projects, workshops, and conferences (DOE, DOI, NEH[11], NSF, SI).

    - Beringia Days Annual Event (DOI's National Park Service; FY2013).

    - Documenting Endangered Languages Program (NEH, NSF; FY2013).

    - Inuit Studies Conference (SI; FY2012).

- Develop tools that Arctic communities can use to more effectively support indigenous languages, traditional ecological knowledge, and natural resource harvesting activities (NEH, NSF, SI; FY2014).

    - Assessing the Vitality of Arctic Indigenous Languages - Research Development Workshops (NSF; FY2012).

    - Recovering Voices Program ongoing outreach (SI; FY2014).

- Create community profiles that highlight continuity of indigenous languages and knowledge systems (NEH, NSF; FY2014).

- Identify and strengthen Federal, State, and local efforts related to indigenous languages, traditional knowledge, and cultural heritage (DOI, NSF, SI; 2014).

- Develop research programs to monitor the status of indigenous languages and knowledge systems and to better understand factors affecting language and cultural resilience (NEH, NSF, SI; FY2016).

Science and Technology Current Gaps and Needs:

Arctic communities have been and will continue to adapt to new technologies, communication, and preservation tools to sustain their cultures and languages. The strengths and threats from the new electronic and media era to indigenous languages have not been adequately studied, nor is a reliable assessment of the opportunities offered by new technologies available to guide Federal and state policies. Digital technologies are being increasingly used in support of indigenous languages and knowledge systems, but many critical issues, such as data management, data sharing, compatibility, and local resources remain unresolved. Research is needed to identify key components necessary for the application of new digital technologies and for collaboration of the many players at the Federal, state, local, and community levels.

---

11. National Endowment for the Humanities

3.7. Understand factors that impact human health in the Arctic, including infectious and non-communicable diseases, climate change, environmental contamination, and behavior and mental-health disorders

Lead Authors:
Alan Parkinson, CDC
Marya Levintova, NIH

Agencies Partners:
CDC, EPA, IHS[12], NIH, USARC

Although health and survival of Arctic indigenous peoples have improved over the past 50 years, important disparities remain in life expectancy, infant mortality, and leading causes of death when compared with their respective national populations. Life expectancy of the indigenous peoples of Alaska, northern Canada, and Greenland is lower than that of the general populations of the United States, Canada, and Nordic countries. Similarly, infant mortality among indigenous segments of those populations is higher than that of the comparable national populations. Meanwhile, mortality rates for heart disease and cancer—once much lower among the indigenous populations of the United States, Canada, and northern European countries—now mirror their respective national rates. In addition, indigenous populations of Alaska, Canada, and Greenland have higher mortality rates for unintentional injury and suicide believed to result from a complex disorder of depression, alcoholism, child abuse, and other substance abuse (Levintova *et al.* 2010). Other health concerns of those groups include a high prevalence of infectious diseases such as hepatitis B and C, *Helicobacter pylori*, infant respiratory syncytial virus (RSV), tuberculosis, and sexually transmitted diseases, as well as heath impacts associated with exposures to environmental pollutants, rapid economic change and modernization, and climate change (Young and Bjerregaard 2008).

The majority of DHHS-supported research in the Arctic is carried out within Alaska in collaboration with scientists at the University of Alaska, the State of Alaska, and the Alaska Native Tribal Health Consortium (ANTHC). Most activities are funded by individual NIH institutes and/or CDC centers; however, some activities receive support from non-DHHS agencies, such as the EPA and U.S. Arctic Research Commission. The CDC maintains the Arctic Investigations Program (AIP), which is a field station located on the ANTHC campus in Anchorage. AIP provides a platform for collaborative biomedical and prevention research focused on improving the health of Alaska Native population and promotes circumpolar health through linkages with CDC centers, the International Union for Circumpolar Health, and the Arctic Council's Sustainable Development Working Group's Arctic Human Health Experts Group (Parkinson 2010). CDC also maintains the Alaska Pacific Regional Office of the National Institute for Occupational Safety and Health, which conducts occupational safety research to reduce hazards in the Arctic.

Supporting further multi-agency research will contribute to knowledge of factors affecting human health in the Arctic and lead to development and implementation of prevention and control and successful treatment strategies that will greatly improve the health and well-being of Arctic residents.

---

12. Indian Health Service

3.7.1 Continue to expand circumpolar surveillance and research for infectious diseases, non-communicable diseases, trauma, injury, sanitation services, and indoor air quality to help prevent morbidity and mortality

Why do this?

Effective surveillance can facilitate timely control of disease outbreaks, inform public health decisions on research and resource allocation, and provide data to maximize prevention and control strategies. Population-based surveillance of diseases of concern is conducted by public-health agencies in the United States, Canada, Greenland, Iceland, Norway, Finland, Sweden, and the Russian Federation. The International Circumpolar Surveillance (ICS) of Emerging Infectious Diseases was established in 2000 as an Arctic Council Sustainable Development Working Group project. Linkage of the individual national systems has created a network of hospitals, public-health agencies, and reference laboratories through-out the Arctic to collect, compare, and share uniform laboratory, epidemiological, and research data on infectious diseases and assist in forming prevention and control strategies (Parkinson *et al.* 2008).

Cancer research depends upon population-based cancer registries for monitoring cancer burden, which can be assessed in terms of mortality, incidence, health-care use, and economic cost. Planning and evaluating cancer screening programs depend upon disease-stage and incidence data provided by population-based registries. At NIH, the National Cancer Institute has supported the Alaska Native Tumor Registry since 1999 as part of its Surveillance, Epidemiology and End Results (SEER) Program. Data from SEER show that gastric cancer is the third most-common type of cancer in the Alaska Native population (Wiggins 2008). Commonly associated with *Helicobacter pylori* infection, the mortality rate for gastric cancer in the Alaska Native people is more than three times that found in the general U.S. population (Miernyk *et al.* 2011). More research is needed to determine the linkage between *Helicobacter pylori* infection and gastric cancer and peptic ulcer disease in Alaska Natives and other indigenous Arctic populations. In addition, there is an urgent need for effective strategies for treating *Helicobacter pylori* in Arctic populations where infection is endemic. Hepatitis B virus infections occur at endemic rates in Arctic populations and result in high rates of long-term problem such as cirrhosis and cancer of the liver. More needs to be learned, however, about the prevalence and clinical impact of hepatitis C (Young and Bjerregaard 2008).

Native people living in Southwest Alaska suffer a high burden of acute and chronic respiratory disease. For example, one in four infants from the region is hospitalized annually with acute respiratory infections. Hospitalization rates of respiratory syncytial virus (RSV) infection in infants are the highest documented (Karron *et al.* 1999). Bronchiectasis, a chronic lung aftermath of severe pneumonias, is common among Alaska Natives in that region (Singleton *et al.* 2000). A combination of substandard housing, overcrowding, poor indoor-air quality, lack of indoor plumbing and running water, and other environmental factors contribute to such conditions (Hennessy *et al.* 2008). Alaska rates last among U.S. states for adequate sanitation service; approximately 23 percent of rural Alaskan households lack in-home water and sewer service. Hospitalization rates for respiratory, skin, and invasive-bacterial infections are all higher among persons living without running water. Research is needed to reduce environmental triggers of respira-tory disease in homes and develop strategies to address the ongoing disparities in sanitation services and the resultant health impacts.

Problems resulting from trauma and injuries are among the most serious that affect people living and working in the Arctic. Injuries are by far the most important causes of death among people around 35 years of age there (Young and Bjerregaard 2008). While unintentional injuries have always been a hazard of living and working in the Arctic, major sociocultural changes and the widespread availability of alcohol have changed the pattern and extent of these injuries. A thorough review of the underlying cause of these health problems (based upon depression and including alcoholism, other drug abuse, child abuse, and suicide) occurred at a meeting in Anchorage on June 2 and 3, 2009, through collaboration between the USARC and the NIH Fogarty Center. That meeting led to a series of recommendations published in Levintova *et al.*, 2010. Alaska also continues to have a high work-related fatality rate, mostly from drownings in the commercial fishing industry and plane crashes in aviation, which are exacerbated by the cold and remote and Alaska climate. More research is needed to understand the causes and potential interventions needed to reduce the morbidity and mortality associated with intentional (suicide) and unintentional injuries. Special attention should continue to be given to preventing drownings and aircraft crashes. Worker health and safety in oil-spill response in the Arctic or other ice-covered waters poses new concerns as interest in natural resource exploration and extraction increase.

Timeframe:

Long-term (5-10 years)

Expected Outcomes:

- The circumpolar surveillance of invasive bacterial diseases (causes of pneumonia, meningitis, septicemia) is ongoing and will be used to monitor the impact of intervention programs as they are implemented.

- Together with the Public Health Agency of Canada, ICS will expand to include surveillance of tuberculosis in the U.S. Arctic, northern Canada, Greenland, and six northern regions of the Russian Federation for the collaborative systematic collection, interpretation, and dissemination of information pertaining to tuberculosis in circumpolar populations for use in epidemiologic study, policy generation, program design, and evaluation.

- Use of cancer registry data will allow planning and evaluation of cancer-screening programs and will contribute to the reduction in cancer incidence and mortality.

- Develop consensus guidelines for the diagnosis and treatment of *Helicobacter pylori* in Arctic populations and undertake a circumpolar study on putative bacterial markers of virulence and host and environmental risk factors associated with peptic ulcer disease and gastric cancer in Arctic indigenous peoples. Potential markers will be validated in a prospective study of gastric cancer patients and matched controls to identify persons at risk for early screening.

- Promote and collaborate on surveillance research and management programs of chronic hepatitis B and C using patient registries, increase screening and vaccination for hepatitis B in Arctic, and design collaborative research programs on virology and pathogenesis of hepatitis B and C in the Arctic.

- A North American Free Trade Agreement funded project will measure air quality and implement home-based intervention strategies to reduce levels of home-based environmental triggers of respiratory disease. Additionally, this project will measure the impact of these interventions on the severity and frequency of respiratory symptoms in Alaska Native children with respiratory disease and their families.

- Improve the water and sanitation services available to rural Alaskans by bringing together various stakeholders for the purposes of determining novel strategies, including single home solutions for water purification and sanitation. The goal is to address the ongoing disparities in sanitation services and the resultant health impacts while keeping the costs of solutions manageable.

- Continued development of strategies to prevent morbidity and mortality among workers in Alaska is needed particularly in the commercial fishing and aviation industries.

- A retrospective study of all Arctic maritime disasters and oil spill is being conducted, which will provide detailed information about casualties sustained during those events as well as hazardous exposures suffered by responders, including search and rescue and spill clean-up personnel. The results of this project may be incorporated into U.S. oil spill contingency plans and could be adopted as an addendum to the revised spill response plan being developed by Emergency Prevention, Preparedness, and Response (EPPR) and/or used as source material for occupational safety and health insertions in that plan.

- Develop strategies to prevent morbidity and mortality among oil spill response workers in the Arctic environment. Results may be incorporated into U.S. oil spill contingency plans, and could be adopted as an addendum to the revised spill response plan being developed by EPPR and/or used as source material for occupational safety and health insertions in that plan.

Milestones:

- Publish a comprehensive report on Circumpolar Surveillance of Infectious Diseases for the Arctic Council Ministerial (CDC; FY2013).

- Conduct 5-year retrospective review of tuberculosis in northern Canada, the U.S. Arctic, Greenland and northern regions of the Russian Federation (CDC; FY2013).

- Maintain Alaska's compliance with standards of the National Program of Cancer registries and the National Association of Central Cancer Registries (CDC; FY2012).

- Draft and publish a *Helicobacter pylori* treatment consensus document for high-prevalence. Validate *Helicobacter pylori* bacterial markers of virulence and host and environmental risk factors associated with peptic ulcer disease and gastric cancer in Arctic indigenous peoples (CDC; FY2013).

- Increase screening and vaccination for hepatitis B in Arctic and the design collaborative research programs on virology and pathogenesis of hepatitis B and C in the Arctic (CDC; FY2012).

- Measure indoor air quality before and after home-based intervention strategies in homes of children with chronic respiratory diseases. Measure severity and frequency of respiratory symptoms in children before and after the interventions (CDC, IHS; FY2013).

- Together with the Alaska Department of Environmental Conservation's Village Safe Water Program conduct stakeholder meetings to address scientific, technological, and policy challenges associated with lack of sanitation services in Alaska. Determine health and economic consequences associated with lack of sanitation services in Alaska using epidemiologic studies and economic models (CDC, USARC; FY2013).

- In collaboration with other Arctic nations and the State of Alaska, publish a supplement to the Arctic Council's *Field Guide for Oil Spill Response in Arctic Waters*, which will contribute understanding and control of the unique hazards that exist to workers responding to an oil spill in the Arctic (CDC, NIOSH[13]; FY2013).

- Maintain the Alaska Occupational Injury Surveillance System in cooperation with the State of Alaska. Publish 20 years of occupational safety research to document progress and set goals for the next 10 years (CDC, NIOSH; FY2014).

- Continue work with the USCG and Native Community Development Groups to identify ways to encourage the use of personal flotation devices and personal locator beacons while commercial fishing. Continue partnerships with marine-safety educators to provide cold-water survival classes in remote Native villages (CDC, NIOSH; FY2014).

- Conduct focus groups with pilots and company owners flying to remote villages to identify strategies to combat pilot fatigue, particularly in the high-risk summer months (CDC, NIOSH; FY2013).

Science and Technology Gaps and Needs:

Lack of sufficient numbers of certified Tumor Registrars has impeded the technological progress of the Alaska Cancer Registry. The absence of electronic means for hospitals and physicians to report to the central cancer registry is a technological gap. Further technological advances are needed to provide in-home sanitation services in Arctic environments. Challenges include providing sufficient clean water for hand washing and cleanliness and sanitation for the approximately 25% of rural Alaskans without these vital services. Novel solutions for single homes of water provision as well as sanitation need to be explored. Other efforts include improving energy efficiency of water systems operations; alternative delivery and waste removal technologies that address specific problems posed by the climate, soil, and permafrost; and changing source water parameters. Technology should be explored that helps achieves a balance between providing adequate volumes of water to maximize healthy washing behavior and conserving water and energy resources.

---

13. National Institute for Occupational Safety and Health (Center for Disease Control)

3.7.2 Continue interagency collaboration to monitor the impacts of climate change and environmental contaminants on human health and wildlife

Why do this?

Over the last three decades, Arctic Alaska's average temperature has increased by 5 degrees F (2.7 degrees C), resulting in extensive glacier melting, thawing permafrost, reduction in Arctic Ocean sea-ice extent, precipitation increases, decrease in duration of snow cover, longer ice-free seasons on lakes, altered food webs, and greater risk of wildland fires. Warmer temperatures may cause environmental contaminants to move more readily from soil and water into the air; foster greater movement of contaminants from lower-latitude source-area soils (industrial sites, agricultural areas, etc.) into Arctic regions; and lead to contamination of traditional foods with man-made chemicals such as polychlorinated biphenyls (PCBs), dioxins, toxophenes, and other pesticides. In the Arctic, those chemicals may be enriched in the marine- and land-based food webs and end up in humans. These compounds may be associated with adverse health outcomes and have caused some Alaska Natives to decrease consumption of traditional marine subsistence species and substitute with less-healthy western foods (ACIA 2005).

The impact of climate change on communities includes the disruption of permafrost-dependent structures, threats to village water supplies and sanitation systems, reduction in availability of subsistence species, and decreased air quality (wildland fire smoke, dust, and pollens). Health threats from climate change include morbidity and mortality resulting from trauma and injury associated with extreme events (storms, floods, increased heat and cold) and unpredictable ice conditions, increased mental and social stress related to changes in environment and loss of traditional lifestyle, decreased access to quality water and food sources, and potential changes in prevalence of some parasitic and zoonotic infectious diseases (Parkinson 2008). Evidence also exists for health consequences from exposure to toxic metals, such as mercury and lead. There may be adverse effects from exposure to organochlorines and mercury on child neuropsychological development and of organochlorines on their immune function and susceptibility to infection.

New and previously experimental or seasonal shipping lanes are being opened in the Arctic as sea ice retreats and oil and mineral exploration markedly increase. These transport and industrial activities raise the possibility of increasing numbers of shipping disasters and oil spills, with increased risk to human workers, as well as to the marine environment, fauna, and flora. The EPPR Worker Hazard Reduction Project will attempt to address some of those emerging hazards.

Consequently, there is need for studies that collect and analyze human-health and ecosystem observations in rural communities located in the many diverse and ecologically-distinct areas of Alaska. In addition, studies that quantify the cumulative health effects of exposure in Alaska Native mothers and their infants to multiple environmental contaminants [anthropogenic persistent organic pollutants (POPs) and mercury] in subsistence foods are needed.

Timeframe:

Mid-term (3-5 years)

Outcomes:

- Community health assessments are designed to assess, one village at a time, the existing and emerging threats and to provide residents with the training and capacity to monitor existing threats, detect new threats, and allow the development of adaptation strategies, as proposed in section 3.6. The village-based monitoring programs conducted by ANTHC will form a network that contributes data to state and Federal agencies to better inform those programs responsible for both human and wildlife health. The outcome of these programs could link with developing programs in Canadian Arctic communities.

- Engaging Alaskans as lay observers of human health and ecological events potentially associated with climate change can provide the public-health community with an important early warning of public-health consequences from such change (in conjunction with efforts in 3.6). A University of Alaska, Anchorage project will provide public-health professionals and other decision-makers with a network for the dissemination of informed and culturally appropriate risk communications to guide adaptation planning that reduces health impacts on the Alaskan population.

- A data base of organochlorines, toxic metals, and micronutrients levels in human and salmon tissue will allow the detection of any associations between prenatal exposure to organochlorines and/or mercury with adverse health outcomes, as well as any positive associations between micronutrients levels and health. A time series in human- and subsistence-species tissue levels of organochlorines and mercury will help evaluate the impact of climate regime change on ocean and atmospheric current delivery of those toxins from lower latitudes into the Bering Sea, addressing possible climate-change impact on subsistence resources and human health. ANTHC will provide and communicate risk and benefit data to Alaska Natives to enable region-wide strategies and policies to reduce the risk and increase benefits of the traditional diet, which will strengthen that critical component of Alaska Native culture.

- Evaluate climate-sensitive infectious-disease surveillance systems for—and baseline levels of infection in—humans and wildlife in Alaska. Such evaluations will result in better understanding of the epidemiology and risk to subsistence wildlife species and to the people who depend of them for food, allowing for rapid detection of outbreaks and development of prevention and control strategies.

Milestones:

- Conduct community health assessments and initiate training and deployment of monitoring technology. Develop a web-based monitoring network and village adaptation strategies, and establish the monitoring database. Conduct statistical analysis of laboratory specimens and share data with agencies and jurisdictions. Provide feedback to tribal leaders. Continue planning with interested agencies to develop support for ongoing monitoring and extension to interested communities in Alaska. Disseminate program results to other circumpolar communities and affected communities in the lower 48 states (IHS; FY2015).

- Develop, deploy, and assess a surveillance and response Toolkit for Alaska to promote community based adaptation planning for climate change. Deliverables will include training protocols, a web portal, a surveillance dataset organized by region, adaptation and mitigation recommendations, and reports/manuscripts for peer-reviewed publication (CDC; FY2013).

- Recruit a cohort of 200 Alaska Native women for collection of blood samples, patient interview, and medical-chart information. Collect and chemically analyze salmon tissue samples (CDC, DOI, EPA; FY2015).

- Conduct surveillance evaluations and sero-prevalence studies on humans and wildlife for potentially climate-sensitive infectious diseases such as those caused by *brucella, trichinella, echinococcus, toxoplasma, francisella, giardia,* and *cryptospordium* species (CDC, DOI; FY2013).

- Implement a NOAA-CDC memorandum of agreement for environmental and public health impacts providing exchange of scientific expertise and resources in the areas of climate, weather, water, and environmental, oceanographic, and atmospheric health as it relates to public health (CDC, DOC; FY2015).

Science and Technology Gaps and Needs:

Findings from community assessments will require engineering solutions to be developed and applied by the ANTHC Division of Environmental Health and Engineering. Technology will need to be developed for ongoing village-based surveillance and determining antibody levels of zoonotic infectious diseases in killed subsistence animals, as well as village-operated sampling for climate-sensitive microbial threats to water security, such as *giardia, cryptosporidium, toxoplasma, tularemia,* and harmful algal blooms. Development of filter-paper blood-spot tests of animal blood are needed to monitor zoonotic diseases and possible contaminants. There is a general need to improve laboratory diagnostics and molecular-typing systems for many potentially climate-sensitive parasitic and zoonotic infectious agents in humans in wildlife. Finally, adaptation and mitigation planning may require local, state, and national coordination to conduct and evaluate further surveillance measures.

3.7.3 Continue to support investigator-initiated research in major health priority areas such as mental health including substance abuse and suicide, obesity, diabetes, and cancer

Why do this?

NIH, the leading U.S. biomedical- and behavioral-research agency, supported more than 80 research projects in the Arctic from 2009-2012, with approximate average annual expenditure of $31 million. These projects are conducted by investigators at individual institutions with many having multiple research partners. NIH funding is awarded primarily to research and related institutions in Alaska and the Arctic, to non-Alaska institutions focused on indigenous populations, and to increase biomedical research capacity among Native and non-Native Alaska populations.

The University of Alaska Center for Biomedical Research Excellence (COBRE) program at the Center for Alaska Native Health Research (CANHR) and the *Alaska Institutional Development Award for Biomedical*

Research Excellence (INBRE) program receive the largest proportion of NIH funds in the region. Both programs are funded through the National Institute for General Medical Studies to support studies of chemical agents (especially contaminants in subsistence foods) and zoonotic and vector-borne microbial agents of disease. CANHR projects represent one example of multi-agency involvement that includes tribal and local communities, CDC, IHS, and other organizations.

NIH-supported projects in the Arctic also focus on behavioral and mental-health problems, including addiction and related disorders. The National Institute on Minority Health and Health Disparities (NIMHD) and the National Institute on Drug Abuse have supported dozens of research projects on the determinants of substance abuse and mental health disorders in the Arctic. Additionally, NCMHD supports studies of various interventions for substance abuse and mental illness in Alaska.

In addition, NIH has engaged with USARC, CDC, the State of Alaska, and local and academic institutions to develop priorities for research increases in Alaska, particularly on health-care delivery innovations, including telemedicine and information technology. The results of a workshop convened by those entities in Alaska indicate there is major need for better communications among agencies in and out of Alaska on issues related to health research, and there is need for inclusion of local communities or tribal groups in research planning and implementation. Inclusion of innovative and culturally appropriate methods of study is also needed (Levintova *et al.* 2010).

Timeframe:

Long-term (5-10 years)

Outcomes:

- NIH will continue to support investigator-initiated, peer-reviewed research projects in biomedical and behavioral sciences in the Arctic. Many NIH-awarded projects engage multiple partners.

Science and Technology Gaps and Needs:

As need for further funding of biomedical and behavioral research in the Arctic continues, NIH will support appropriate investigator-initiated, peer-reviewed projects. A number of gaps, however, should be highlighted:

- Alaska-based institutions and researchers have not historically been highly competitive peer review. In this time of reduced economic constraints, this gap is difficult to fill. Therefore, there is a substantial need to develop collaborations with successful NIH award recipients.

- It is difficult to recruit from the sparsely populated Native villages in Alaska the large numbers of participants needed to conduct longitudinal studies of chronic diseases. Genetic and cultural similarities among Arctic native populations would allow for statistically meaningful cohorts if communication and information exchange among researchers were optimal.

- Finally, in addition to large population-based studies, exploration of alternative research design (such as small sample design and studies in low-resource settings) is needed.

3.7.4 Continue to engage indigenous communities and tribal groups in research activities and projects in the Arctic

Why do this?

Indigenous community members and their leaders should be involved in all stages of the research process—from formulating and approving projects and methods, to determining research outcomes, to interpreting and disseminating results. Genuine collaboration between researchers and indigenous communities not only builds valuable partnerships and mutual trust, but it also enhances research through information and knowledge sharing. Optimal engagement and partnership efforts enable research that is both culturally sensitive and conducive to producing shared benefits.

Timeframe:

Long-term (5-10 years)

Outcomes:

- NIH, NSF, and CDC-supported investigators, and others, interested in conducting research in the Arctic/Alaska work closely with community and tribal organizations/leaders in developing and conducting their research. In many cases, projects require tribal permission to be carried out in the community.

- National Heart Lung and Blood Institute provided funding to the Norton Sound Health Corporation, an Alaska Native-owned health corporation, as one of three grantees to conduct the second phase of the Genetics of Coronary Artery Disease in Alaska Natives study.

Milestones:

- NIH and CDC supports research through CANHR where researchers collaborate with tribal and local communities on a number research topics, including nutrition, alcohol consumption and abuse prevention, drug abuse and prevention, and others (NIH; FY2012).

Science and Technology Gaps and Needs:

Better communication and more information exchanges are necessary between the researchers and the community before the start of research, during, and after its completion.

References

ACIA. (2008). Arctic Climate Impact Assessment Report. Cambridge: Cambridge University Press UK. p863-906.

ACIA. (2005). Arctic Climate Impact Assessment: Scientific Report. 1042 pp. Cambridge University Press, UK.

Bengtsson et al., 2005. Second International Conference on Arctic Research Planning, Science Plan 9, Copenhagen, Denmark. http://www.icarp.dk.

Federal Actions for a Climate Resilient Nation: Progress Report of the Interagency Climate Change Task Force. October 28, 2011. http://www.whitehouse.gov/sites/default/files/microsites/ceq/2011_adapta-tion_progress_report.pdf.

Hennessy, T.W., Ritter, T., Holman, R., Bruden, D., Yorita, K., Bulkow, L.R., et al. (2008). The relationship between in-home water service and the risk of respiratory tract, skin and gastrointestinal tract infections among rural Alaska Natives. *American Journal of Public Health*, 98(11): 2072-2078.

Karron, R., Singleton, R., Bulkow, L., et al. (1999). Severe respiratory syncytial virus disease in Alaska Native children. *J Infect Dis* :180, 41-9.

Levintova, M., Zapol, W.I., Engmann, N. (Eds.). (2010). Behavioral and Mental Health Research in the Arctic: A Strategy Setting Meeting. *Inter J. Circumpolar health Suppl. 5.* http://www.ijch.fi/CHS/CHS_2010(5).pdf.

Miernyk, K., Morris, J., Bruden, D., McMohn, B., Hurlburt, D., Sacco, F., Parkinson, A.J., Hennessy, T., & Bruce, M. (2011). Characterization of helicobacter pylori cagA and vacA genotypes among Alaskans and their correlation with clinical disease. *J.Clin.Micro* 49(9), 3114-3121.

Parkinson, A. J. (2010). Improving human health in the Arctic: the expanding role of the Arctic Council's sustainable development working group. *International J. Circumpolar Health* 69(3), 304-313.

Parkinson, A. J., Bruce, M. G., & Zulz, T. (2008). International Circumpolar Surveillance, and Arctic network for the surveillance of infectious diseases. *Emerg. Infect. Dis.* 14(1), 18-24.

Parkinson, A. J. (2008). Climate Change and Infectious Diseases: the Arctic Environment. IOM Institute of Medicine. Global climate change and extreme weather events: understanding the contributions to infectious disease emergence. Washington DC: The National Academies Press.

Roberts, A. and coauthors (2010). A Science Plan for Regional Arctic System Modeling, A report to the National Science Foundation from the International Arctic Science Community. International Arctic Research Center Technical Papers 10-0001. International Arctic Research Center, University of Alaska Fairbanks.

Wiggins, C. L., Predue, D. G., Henderson, J. A., Bruce, M., Lanier, A. P., Kelly, J. J., Seals, B. F., & Espey, D. K. (2008). Gastric Cancer among American Indians and Alaska Natives in the United States 1999-2004. *Cancer*, 113 (5), 1225-1233.

Singleton, R., et al. (2000). Bronchiectasis in Alaska Native children: causes and clinical courses. *Pediatric Pulmonology*, 29(3): 182-187.

Young, K. T, & Bjerregaard, P. (2008). *Health Transitions in Arctic populations*. Toronto: *University of Toronto Press.*

Chapter 4: Research Infrastructure

Because the Arctic is geographically remote and environmentally harsh, advancing regional knowledge and understanding requires specialized research platforms and instruments. Needed infrastructure ranges from direct on-the-ground observations to satellite observations from space with advanced instruments and from field stations to research vessels. National and international assets are regularly brought to bear. United States infrastructure, its use, and availability are summarized in Tables 1-3. A few relevant foreign assets and missions are also mentioned, but the list is not exhaustive. International coordination of infrastructure and cost sharing is highly desirable.

Table 1 summarizes major infrastructure, including space- and ocean-based assets, aircraft (piloted and unpiloted), and field stations. Space-based instruments operated by NASA, NOAA, and other agencies are especially powerful tools for observing the remote Arctic. Reliance on foreign satellites for some observations and measurements is now common. NASA and its partners, including some commercial entities, operate nearly 60 research aircraft, many of which apply to Arctic work. The aircraft most critical to this research plan are listed in Table 1.

Research in the Arctic Ocean and adjacent seas employs ice-breaking or ice-strengthened vessels, which are operated by the Coast Guard, the University-National Oceanographic Laboratory System (UNOLS), NOAA, private contractors, and foreign governments. U.S. Navy submarines that deploy occasionally to the Arctic Ocean obtain a limited amount of research-quality data. Sea-ice camps are occasionally operated for brief periods in the Arctic Ocean for process studies and autonomous instrument deployment. As sea-ice cover in the Arctic Ocean diminishes and vessel traffic increases, national needs for Arctic vessels for research, national security, and marine safety are being assessed.

Field stations on land and the Greenland Ice Sheet are especially useful for long-term observations and for supporting process studies and shorter-term, satellite field camps.

Table 2 summarizes significant but smaller research tools, with a focus on airborne and ocean-based infrastructure. The latter include numerous autonomous platforms and instruments that are growing in importance as access to research vessels becomes more challenging and the need for year-round observations grows. Basic equipment to enhance indigenous observations is also included.

Table 3 summarizes key assets of cyberinfrastructure that are important for research and education in the Arctic. Electronic media and distribution systems are critical to data acquisition, transmission, archiving and distribution, and to communicate research results to the greater scientific community and public. As data on the Arctic environmental system has become more voluminous, the need for more sophisticated hardware, software, standards, and sharing agreements has increased.

Table 1. Major U.S. infrastructure (space-based, aircraft, ocean-based, field stations) needed to accomplish the five-year Arctic research plan. For each infrastructure element, its use, availability, and relevant sections of the plan are identified.

INFRASTRUCTURE	USE	AVAILABILITY	SECTION
Space-based			
Existing satellite missions critical to Arctic research			
NOAA satellite missions	Weather and key climate variables.	Available through 2017.	3.1-3.4
Defense Meteorological Satellite Program (DMSP)	Mapping sea ice with passive microwave.	Available through 2017.	3.1-3.4
NASA Earth Observing Satellites	Detailed studies of sea ice, clouds, and other Arctic parameters.	Many are past design life.	3.1-3.4
Joint Polar Satellite System (JPSS)	Next-generation weather satellite.	SUOMI-NPP has planned operational life to 2017; other satellites are in planning stages.	3.1-3.4
USGS Landsat-5 and -7	Agriculture, geology, forestry, regional planning, mapping, global change research, emergency response and disaster relief, education.	Landsat-5 launched in 1984 and still in operation, but data acquisition limited by an electronics problem. Landsat-7 launched in 1999 and still in operation. Minimum design life of 5 years.	3.1-3.4
SAR (Synthetic Aperture Radar)	Sea ice and glacier geophysics and mapping; Marine transportation support; Oceanography; Mapping—vegetation, geology, topography.	No U S. SAR instruments available. Foreign SAR data (e.g., RADARSAT, TerraSAR-X, COSMO SkyMed) are available for purchase.	3.1-3.4
Satellites planned for launch by 2017			
USGS/NASA LandSat Data Continuity Mission (LDCM)	Agriculture, geology, forestry, regional planning, mapping, global change research, emergency response and disaster relief, education.	Launch in 2013.	3.1-3.4
NASA Global Precipitation Measurement (GPM)	Measure snowfall and heavy rain.	Launch in 2014; Limited footprint over polar regions.	3.3
NASA/DLR (Germany) Gravity Recovery and Climate Experiment (GRACE) follow-on	Arctic oceanography, changes in ice mass, terrestrial water storage.	Launch in 2017.	3.1-3.4
NASA Soil Moisture Active Passive (SMAP)	Soil moisture, freeze thaw patterns, and potentially sea-ice mapping.	Launch in 2015.	3.1-3.4
NASA ICESat 2	Altimetry over land and sea ice to measure changes in thickness.	Launch in 2016.	3.1-3.4

INFRASTRUCTURE	USE	AVAILABILITY	SECTION
Aircraft			
Piloted aircraft			
P-3 (NASA)	Two long-range survey aircraft that can be outfitted with a variety of instruments. Based at NASA Wallops. Currently used for study of Arctic sea ice and land ice, but has applications to many areas of Arctic science.	Available through 2017.	3.1, 3.4
DC-8 (NASA)	One long-range survey aircraft that can be outfitted with a variety of instruments Based at NASA Dryden. Currently used for study of Arctic sea ice and land ice, but has applications to many areas of Arctic science.	Available until 2014, when the aircraft requires major refurbishment.	3.1, 3.3, 3.4
Gulfstream-V (NSF/NCAR)	One long-range, high-altitude aircraft for study of Chemistry and Climate, Chemical Cycles, Air Quality, Mesoscale Weather, Upper Troposphere and Lower Stratosphere.	Available through 2017.	3 3
LC130 (NSF)	Eight ski-equipped heavy-lift aircraft for snow and ice landings to support field operations in Greenland. Based at Stratton Air Base.	Available through 2017.	Indirectly 3 2, 3 3, 3.4
C130Q (NSF/NCAR)	One long-range, high-altitude aircraft with standard thermodynamic, microphysics and radiation sensors, and ability to carry a wide variety of scientific payloads.	Available through 2017.	3 3
C130 (USCG)	Multiple aircraft supporting Homeland Security missions, maritime/Arctic domain awareness, and testing capabilities of personnel and equipment. Scientists and scientific instruments are also accommodated. Based at Coast Guard Air Station Kodiak.	Available through 2017.	3.1, 3.3, 3.4

INFRASTRUCTURE	USE	AVAILABILITY	SECTION
Twin Otter	NOAA. Four aircraft in the fleet. In the Arctic, aircraft are used for marine mammal surveys in the Bering, Chukchi and Beaufort seas.	Available through 2017.	3.1
Small aircraft	USFWS (DOI) annually contracts light aircraft for fish surveys.		3.2
Unpiloted aircraft			
Global Hawk (NASA and NOAA)	Multiple aircraft supporting atmospheric measurements from high altitude (65,000 feet). Based at NASA Dryden.	Available through 2017. Two aircraft are operational for research, others could be made available.	3.1, 3 3, 3.4
Ikhana (NASA)	One aircraft being developed for sea ice and oceanographic measurements, including buoy drop capability. Based at NASA Dryden.	In development for availability in 2013; availability beyond 2013 dependent on research use.	3.1, 3.4
Other unpiloted aerial vehicles	Various medium and small UAS and facilities are being developed for physical, chemical, and biological measurements by DOE, NOAA and NSF. DOE is developing a UAS operations facility at Oliktok (Alaska) site by mid-2012.	Available through 2012.	
Ocean-based			
Icebreakers	Physics, chemistry and biology of ice, ocean and atmosphere; Marine geophysics and bathymetry.	The U S. Coast Guard has three polar icebreakers: *Healy, Polar Star, Polar Sea. Healy* remains active with an expected service life to 2030. *Polar Star* is undergoing an extensive overhaul intended to lengthen its service life to approximately 2020-2021 and is expected to be back in service in FY 2013. It is expected that *Polar Sea* will be decommissioned within the year.	3.1, 3 3, 3.4
Ice-capable research vessels	Physics, chemistry and biology of ice, ocean and atmosphere; Marine geophysics and bathymetry.	The UNOLS vessel *R.V. Sikuliaq* is scheduled to be available for research in early 2014.	3.1, 3 3, 3.4

INFRASTRUCTURE	USE	AVAILABILITY	SECTION
Other U.S. research and survey vessels	Physics, chemistry and biology of ice, ocean and atmosphere; Marine geophysics and bathymetry.	The UNOLS vessel *R.V. Marcus G. Langseth* supported Arctic Ocean geophysical research in summer 2011, demonstrating potential for future use of non-ice strengthened UNOLS vessels. NOAA seasonally deploys the Oscar Dyson (fishery survey class) and hydrographic survey vessels, *Rainier* and *Fairweather*. USFWS (DOI) *R.V. Tiglax* operates in Arctic waters each summer.	3.1, 3.3, 3.4
Foreign vessels	Physics, chemistry and biology of ice, ocean and atmosphere; Marine geophysics and bathymetry.	NOAA annually charters the *R.V. Khromova* from Russia to carry out the Russian-American Long-term Census of the Arctic—servicing a Bering Strait mooring array and observing climate impacts on sea ice, ocean conditions, and ecosystems. U.S. researchers regularly work from the Canadian Coast Guard icebreakers *CCGS Louis St. Laurent* and *CCGS Wilfred Laurier.* International partners (Korea, China, Japan, Russia, Canada) use their vessels to sample in the Distributed Biological Observatory (DBO).	3.1, 3.3, 3.4
Submarines	Physics, chemistry and biology of ice, ocean and atmosphere; Marine geophysics and bathymetry.	Occasional (1-3/year) U.S. Navy Science Accommodation Missions (SAMs). Also occasional (every few years) Royal Navy (UK) submarine missions.	3.1, 3.4, 3.5

INFRASTRUCTURE	USE	AVAILABILITY	SECTION
Sea ice	Physics, chemistry and biology of ice, ocean and atmosphere; Marine geophysics and bathymetry.	Ephemeral facilities that are occupied typically for periods of weeks to months, and less commonly for 1-2 years.	3.1, 3 3, 3.4
Field Stations			
Toolik Lake, North Slope of Alaska	Primarily investigator-led biology and long-term ecological research.	Operations supported by NSF, research supported by NSF and others, and available through 2017.	3.2, 3.4
Summit, Greenland	Primarily investigator-led snow and ice geophysics, atmospheric science, and long-term observing. Also host to an emerging NOAA Baseline Observatory for long-term monitoring of atmospheric constituents.	Operations supported by NSF and NOAA, research supported by NSF and others, and available through 2017.	3.2, 3 3, 3.4
Barrow, Alaska	NOAA Baseline Observatory for long-term monitoring of atmospheric constituents. DOE maintains an Atmospheric Radiation Measurement (ARM) research site at Barrow, and a new site at Oliktok, Alaska, will be operational mid-2013. The Next Generation Ecosystem Experiment is expected to operate at Barrow and on the Seward Peninsula, Alaska, beginning in 2012.	Available through 2017.	3.2, 3 3, 3.4
Tiksi, Russia	Primarily atmospheric science.	NSF maintains the observatory in cooperation with NOAA and the Russian Federal Service for Hydrometeorology and Environmental Monitoring.	3.3, 3.4

Table 2. Significant smaller airborne and ocean-based assets and instruments needed to accomplish the five-year Arctic research plan. For each infrastructure element, its use, availability, and relevant sections of the plan are identified.

INFRASTRUCTURE	USE	AVAILABILITY	SECTION
Airborne			
LIDAR for ice altimetry	Sea ice thickness. Land ice surface elevation change. NASA Goddard provides various LIDARs for Arctic work for piloted and unpiloted aircraft.	Available through 2017.	3.1, 3.4
Radar, ice penetrating	Snow thickness over sea ice. Bed maps under land ice. Structure of land ice. NASA and NSF support various institutions that develop and operate snow and ice radars.	Available through 2017.	3.1, 3.4
Ocean-based			
Sea ice-based autonomous observatories (e.g., ice-tethered profilers, flux buoys, ice mass balance buoys, wave buoys)	Physics, chemistry and biology of ice, ocean and atmosphere.	Available, but development of new and improved sensors continues.	3.1, 3.3, 3.4, 3.5
Open-ocean autonomous platforms, (e.g., buoys, wavegliders, wave buoys)	Physics, chemistry and biology of ice, ocean and atmosphere.	Available, but development of new and improved sensors and platforms continues.	3.1, 3.3, 3.4, 3.5
Unmanned underwater vehicles (UUV)	Physics, chemistry and biology of water; sea-ice draft; bathymetry.	Short duration: available. Long duration: not available, but under development.	3.1, 3.4, 3.5
Acoustic communication and navigation	Command and control, and data relay to shore, for UUVs and similar platforms.	Not available, but technology exists.	3.1, 3.4
Moorings	Physics, chemistry and biology of water; sea-ice draft.	Available, but development of new and improved sensors continues.	3.1, 3.4, 3.5
Cabled ocean observatories	Physics, chemistry and biology of near-shore waters and ice, and real-time transmission of data; acoustic communication and navigation nodes.	With one exception in the Canadian Arctic, not available.	3.1, 3.4, 3.5
Other			
Navigation (e.g., GPS) and communication tools (e.g., radios)	Technological innovations to enhance the collection of indigenous observations.	Available, but not widely distributed specifically for research-related observations.	3.1, 3.2, 3.4, 3.6, 3.7

Table 3. Cyberinfrastructure needs for the five-year Arctic research plan. For each element, its use, availability, and relevant sections of the plan are identified.

CYBERINFRASTRUCTURE	USE	AVAILABILITY	SECTION
Broadband communications and Internet connectivity	Essential components of cyberinfrastructure for research and education.	Broadband communications and connections to the Internet remain a challenge in large areas of the Arctic, thus placing limits on movement of and access to data and information, and education and outreach opportunities for northern residents.	Applies to all sections.
Virtual research networks	Facilitate online communication, collaboration and scientific discovery among researchers.	The NSF Research Coordination Network (RCN) Program addresses the need for the development of virtual research networks. The DOD-funded Arctic Collaborative Environment (ACE) potentially offers a similar capability, with an emphasis on Web-based visualization, analysis and interpretation of diverse datasets.	Applies to all sections.
Data archiving	Long-term data curation and distribution services.	Available for many data sets, e.g , NASA EOSDIS, NOAA National Data Centers (NGDC, NODC, NCDC), DOI/USGS EROS Center, DOE ARM and NGEE data archives. Arctic-relevant data from global and regional model simulations (e.g., Earth System Grid Federation). NSF Division of Arctic Sciences funds the development of Advanced Cooperative Arctic Data and Information System (ACADIS) for research data and Exchange for Local Observations and Knowledge (ELOKA) in the Arctic for traditional knowledge. Inter-operability remains a challenge. Human subjects data and traditional knowledge require special handling due to privacy and data ownership concerns.	Applies to all sections.

CYBERINFRASTRUCTURE	USE	AVAILABILITY	SECTION
Data standards	Adherence to standards from the moment data are acquired makes possible inter-operability of archives, data discovery, analysis and integration, and derived product development.	Data standards are in widespread use, but are not necessarily uniform; different standards are used by different disciplines and research communities. Integrated, multi-disciplinary data standards will be needed for those archives that curate and distribute diverse datasets. As it continues to take on a greater role in responsible data and information management, the scientific community is developing and adopting data standards in consultation with data managers. The NSF Coordination Networks (RCN) program encourages RCNs that will develop community standards for data and meta-data.	Applies to all sections.
Data preparation and processing	QA/QC, preparation of data documentation, metadata profiles and data files, and placement in archives.	Accomplished with varying degrees of quality and success. There is growing recognition that research data must be properly processed and archived. NSF data policy provides guidance, requirements, standards.	Applies to all sections.
Data agreements and access	Facilitate free and open data sharing and exchange nationally and internationally.	In the United States, data are generally freely and openly available, but delays can occur between acquisition and availability in archives. Internationally, the situation is more variable; some countries charge fees and some do not release data.	Applies to all sections.
Data exfiltration	Timely transfer of data from instruments in the field to people for data processing.	Time varies according to circumstances. There is increasing use of satellite communications, e.g., Iridium, for transfer from remote surface instruments. Not available for instruments below the surface leading to delays in data recovery and subsequent broader availability. Broadband communications, including microwave-enabled broadband, remain a challenge in the Arctic—see also 'Broadband communications and Internet connectivity'.	Applies to all sections.

Chapter 5: Acronyms

Acronym	Full Name
A-CADIS	Advanced Cooperative Arctic Data Information Service
ACIA	Arctic Climate Impact Assessment
ADIWG	Alaska Data Integration Working Group
AEC	Atomic Energy Commission
AeroCan	AERosol CANada
AERONET	Aerosol Robotic Network
AIP	Arctic Investigations Program
AIRS	Atmospheric Infrared Sensor
AMAP	Arctic Monitoring and Assessment Program working group
ANTHC	Alaska Native Tribal Health Consortium
AOMIP	Arctic Ocean Model Intercomparison Project
AON	Arctic Observing Network
AOOS	Alaska Ocean Observing System
ARCPAC	Aerosol, Radiation, and Cloud Processes Affecting Arctic Climate
ARCTAS	Arctic Research on the Composition of the Troposphere from Aircraft and Satellites
ARCUS	Arctic Research Consortium of the United States
ARM	Atmospheric Radiation Measurement
ARPA	Arctic Research and Policy Act of 1984
BOEM	Bureau of Ocean Energy Management
BSSN	Bering Sea Sub-Network
CADIS	Cooperative Arctic Data and Information Service
CAFF	Conservation of Arctic Flora and Fauna working group
CALIOP	Cloud-Aerosol Lidar with Orthogonal Polarisation
CANHR	Center for Alaska Native Health Research
CBMP	Circumpolar Biodiversity Monitoring Program
CCGS	Canadian Coast Guard Ship
CCIN	Canadian Cryospheric Information Network

CDC	Centers for Disease Control and Prevention
CESM	Community Earth System Model
CLIVAR	World Climate Research Programme's Climate Variability and Predictability Program
COBRE	Center for Biomedical Research Excellence (University of Alaska)
CSA	Canadian Space Agency
DBO	Distributed Biological Observatory
DEM	Digital Elevation Model
DHHS	Department of Health and Human Services
DHS	Department of Homeland Security
DOC	Department of Commerce
DOD	Department of Defense
DOE	Department of Energy
DoEd	Department of Education
DOI	Department of the Interior
DOS	Department of State
DOT	Department of Transportation
EALÁT	Reindeer Herders Vulnerability Network Study
EC	Environment Canada
ELOKA	Exchange for Local Observations and Knowledge
EOSDIS	Earth Observing System Data and Information System
EPA	Environmental Protection Agency
EPPR	Emergency Prevention, Preparedness and Response Working Group
EROS	Earth Resources Observation and Science
ESA	European Space Agency
ESMF	Earth System Modeling Framework
ESMs	Earth System Models
FMI	Finnish Meteorological Institute
FWS	Fish and Wildlife Service
GAO	Government Accountability Office
GCMs	General Circulation Models

GINA	Geographic Information Network (University of Alaska)
GIS	Geographic Information Systems
GPS	Global Positioning System
GRACE	Gravity Recovery and Climate Experiment
GRISO	Greenland Ice Sheet and Ocean working group
IARPC	Interagency Arctic Research Policy Committee
IASOA	International Arctic System for Observing the Atmosphere
ICARP II	Second International Conference on Arctic Research Planning
ICESat	Ice, Cloud and land Elevation Satellite
ICS	International Circumpolar Surveillance
IHS	Indian Health Service
INBRE	Institutional Development Award for Biomedical Research Excellence
INTERACT	International Network for Terrestrial Research and Monitoring in the Arctic
IOM	Institute of Medicine
IPCC	Intergovernmental Panel on Climate Change
IPCC AR4	Intergovernmental Panel on Climate Change Fourth Assessment Report
IPY	International Polar Year
ISDAC	Indirect and Semi-Direct Aerosol Campaign
JPSS-1	Joint Polar Satellite System-1
LCCs	Landscape Conservation Cooperatives
LIDAR	Light Detection And Ranging
LTER	Long Term Ecological Research
MISR	Multi-angle Imaging SpectroRadiometer
MIZOPEX	Marginal Ice Zone Observations and Processes Experiment
MMC	Marine Mammal Commission
MODIS	Moderate Resolution Imaging Spectroradiometer
NASA	National Aeronautics and Space Administration
NCDC	National Climatic Data Center
NCMHD	National Center for Minority Health and Disparities
NEH	National Endowment for the Humanities

NGDC	National Geophysical Data Center
NGEE	Next-Generation Ecosystem Experiments
NGO	Non-governmental Organization
NIH	National Institutes of Health
NIOSH	National Institute for Occupational Safety and Health
NMFS-AFSC	National Marine Fisheries Service—Alaska Fisheries Science Center
NOAA	National Oceanic and Atmospheric Administration
NODC	National Oceanographic Data Center
NPP	National Polar-orbiting Partnership
NPS	National Park Service
NSF	National Science Foundation
NSIDC	National Snow and Ice Data Center
NSSI	North Slope Science Initiative
NWS	National Weather Service
ONR	Office of Naval Research
PAG	Pacific Arctic Group
PCBs	Polychlorinated Biphenyls
PI	Principal Investigator
POPs	Persistent Organic Pollutants
QA/QC	Quality Assurance/Quality Control
RADARSAT	Radar Satellite
RSV	Respiratory Syncytial Virus
RV	Research Vessel
SAON	Sustaining Arctic Observing Networks
SBIR	Small Business Innovation Research
SDWG	Sustainable Development Working Group
SEARCH	Study of Environmental Arctic Change
SEER	Surveillance, Epidemiology and End Results
SI	Smithsonian Institution
SIKU	Sea Ice Knowledge and Use

SLCFs	Short-lived Climate Forcers
SMAP	Soil Moisture Active Passive
SS	Sea Surface
SSH	Sea Surface Height
SST	Sea Surface Temperature
SWIPA	Snow, Water, Ice, and Permafrost in the Arctic
TES	Tropospheric Emission Spectrometer
UAS	Unpiloted Aerial Systems
UAV	Unmanned Aerial Vehicle
UNOLS	University-National Oceanographic Laboratory System
USARC	United States Arctic Research Commission
USCG	United States Coast Guard
USGS	United States Geological Survey
UUV	Unmanned Underwater Vehicle
VIIRS	Visible Infrared Imager Radiometer Suite
WCRP	World Climate Research Program
WMO	World Meteorological Organization